Praise for Quality Function Deployment

The author is a recognized expert in the field and has plenty of experience in applications. This is well reflected in the quality of writing.

> Madhav S. Phadke
> President
> Phadke Associates, Inc.

There are no books currently available that really address the need for a clear and easy-to-read text on how you actually go about doing QFD. This book is excellent and fills that gap. I have been struggling with some very real situations, this book has given me some new insights on them! I am strongly recommending this book to all of my colleagues and students.

> Kathy Butler
> Engineering Specialist
> Rocketdyne Division
> Rockwell International

Mr. Cohen's book, *Quality Function Deployment: How to Make QFD Work for You*, is excellent for the QFD novice, as well as the practitioner. Mr. Cohen does an excellent job of explaining the QFD process, including the subtle nuances of the methodology. The book is especially strong in linking the marketing activities to the engineering activities.

> Steve Ungvari
> President
> ASI International
> American Supplier Institute

The author combines ten years of personal QFD experience with insights gained from a thorough study of various approaches to create a well-rounded book.

> Bob King
> Chairman and CEO
> GOAL/QPC

Lou Cohen has done a splendid job in documenting the QFD Process in the most practical of manners. I am sure the book will become popular. You at Addison-Wesley are to be commended for publishing such a fine addition to the bookshelves of Quality.

> Tony Hutchings
> Digital Equipment Corporation

QUALITY FUNCTION DEPLOYMENT

ENGINEERING PROCESS IMPROVEMENT SERIES

John W. Wesner, Ph.D., P.E., Consulting Editor

QUALITY FUNCTION DEPLOYMENT

How to Make QFD Work for You

Lou Cohen

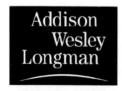

An imprint of Addison Wesley Longman, Inc.

Reading, Massachusetts • Harlow, England • Menlo Park, California
Berkeley, California • Don Mills, Ontario • Sydney
Bonn • Amsterdam • Tokyo • Mexico City

The publisher offers discounts on this book when ordered in quantity for special sales.

For more information, please contact:

Corporate & Professional Publishing Group
Addison-Wesley Publishing Company
One Jacob Way
Reading, Massachusetts 01867

Library of Congress Cataloging-in-Publication Data

Cohen, Lou, 1937–
    Quality function deployment : how to make QFD work for you / Lou Cohen.
       p.   cm. — (Engineering process improvement series)
    Includes bibliographical references and index.
    ISBN 0-201-63330-2 (acid-free paper)
    1. Production planning,   2. Quality function deployment.
I. Title.   II. Series.
TS176.C58 1995
658.5—dc20
                                         95-9903
                                           CIP

Text printed on recycled and acid-free paper.
ISBN 0-201-63330-2

13 PHX 04

Engineering Process Improvement Series

Consulting Editor, John W. Wesner, Ph.D., P.E.

Global competitiveness is of paramount concern to the engineering community worldwide. As customers demand ever-higher levels of quality in their products and services, engineers must keep pace by continually improving their processes. For decades, American business and industry have focused their quality efforts on their end products rather than on the processes used in the day-to-day operations that create these products and services. Experts across the country now agree that focusing on continuous improvements of the core business and engineering processes within an organization will lead to the most meaningful, long-term improvements and production of the highest-quality products.

Whether your title is researcher, designer, developer, manufacturer, quality or business manager, process engineer, student, or coach, you are responsible for finding innovative, practical ways to improve your processes and products in order to be successful and remain world-class competitive. The **Engineering Process Improvement Series** takes you beyond the ideas and theories, focusing in on the practical information you can apply to your job for both short-term and long-term results. These publications offer current tools and methods and useful how-to advice. This advice comes from the top names in the field; each book is both written and reviewed by the leaders themselves, and each book has earned the stamp of approval of the series consulting editor, John W. Wesner.

Key innovations by industry leaders in process improvement include work in benchmarking, concurrent engineering, robust design, customer-to-customer cycles, process management, and engineering design. Books in this series will discuss these vital issues in ways that help engineers of all levels of experience become more productive and increase quality significantly.

All of the books in the series share a unique graphic cover design. Viewing the graphic blocks descending, you see random pieces coming together to build a solid structure, signifying the ongoing effort to improve processes and produce quality products most satisfying to the customer. If you view the graphic blocks moving upward, you see them breaking through barriers—just as engineers and companies today must break through traditional, defining roles to operate more effectively with concurrent systems. Our mission for this series is to provide the tools, methods, and practical examples to help you hurdle the obstacles, so that you can perform simultaneous engineering and be successful at process and product improvement.

The series is divided into three categories:

Process Management and Improvement   This includes books that take larger views of the field, including major processes and the end-to-end process for new product development.

Improving Functional Processes   These are the specific functional processes that are combined to form the more inclusive processes covered in the first category.

Special Process Topics and Tools   These are methods and techniques that are used in support of improving the various processes covered in the first two categories.

This book is dedicated to the memory of my father, who did not get to see its completion, to my mother, who encouraged me to be true to myself, and to my wife, Jane, who helped me in countless ways. Most important, she told me how great this book is, even though it isn't.

Contents

Foreword

Lou Cohen has written the definitive book on quality function deployment (QFD). Every product person needs to have ready access to these best practices. The activities needed to transform QFD from the beautiful abstract concept that has attracted executive attention into work processes that provide customer satisfaction are described in operational detail.

For the last 15 years there has been a great emphasis on the customer viewpoint. However, most product people have been uncertain of the approach to be taken. Based on his great personal experience and insight, Lou's description of the key process of obtaining the voice of the customer will greatly enrich your practice. Applications of QFD bring great benefit to a diverse range of "products," including hardware, software, and services.

In 1984 I started advocating the use of QFD at a time when only a handful of people in North America and Europe had heard of it. In 1988 John Hauser and I wrote an article for the *Harvard Business Review* which attracted great attention. However, the great interest in QFD has not always translated into successful practice. Although QFD looks deceptively simple, success requires subtle skills that go beyond ad hoc approaches. This book will help you make a strong start on the journey to master those subtle skills. QFD provides integration along the value stream to achieve market success. Use this book creatively, and your reward will be great.

Don Clausing
Cambridge, Massachusetts
January 1995

Dame Nature gives her orders to Genius*
An early depiction of the process of gathering
product requirements.

*Roman de la Rose, Flemish, about AD 1500, Harley MS.
4425,f.166, by permission of The British Library

Introduction

Over the last twenty years, companies in the United States have moved toward new styles of doing business, based on overseas competitive pressures, the needs of global economics, and the advances of technology.

U.S. companies have taken many steps to become more competitive. Among them has been the adoption of the Total Quality Management (TQM) approach or one of its many aliases, all of which have stressed customer-driven planning, continuous improvement, and employee empowerment.

A key component of TQM is the adoption of "tools" to assist in creative thinking and problem solving. These are not physical tools, such as computers or micrometers; instead, they are *methods* that relate ideas to ideas, ideas to data, and data to data; that encourage team members to communicate more effectively with each other; and that help teams to effectively formulate business problems and their solutions.

Quality Function Deployment (QFD) is an adaptation of some of the TQM tools. In Japan, in the late sixties, QFD was invented to support the product design process (for designing large ships, in fact). As QFD itself evolved, it became clear to QFD practitioners that it could be used to support service development as well.

Today, its application goes considerably beyond product and service design, although those activities are quite commonly supported by QFD. QFD has been extended to apply to any planning process where a team has decided systematically to prioritize their possible responses to a given set of objectives. The objectives are called the "Whats,"[1] and the responses are called the "Hows." QFD provides a method for evaluating "How" a team should best accomplish the "Whats."

I first learned about QFD in January, 1986. Since then I have used QFD—and helped others use it—for software and hardware product design, service development, and many other special applications involving creative interpretations of the "What" and the "How" of QFD. You may assume that because I have written this book I must be enthusiastic about QFD. In fact, I have used QFD for my own projects, and I have helped other development and service groups use QFD on almost a continuous basis since those early months of 1986. I am continually amazed and delighted at the power and robustness of the method. Teams I work with are delighted with the results of QFD, because the process of using QFD helps the team ask the crucial questions whose answers will give them a highly competitive product or service.

Despite my enthusiasm for QFD, I have attempted to keep the hype out of this book. I am personally turned off by a hard sell—I like to gather the facts and make my own decisions. Therefore, except for a few instances of irrepressible enthusiasm, my style in this book has been mainly to present the information and let you, the reader, decide what's best for you. As for those readers who want reassurance that they are not wasting their time reading this book—let me tell you, QFD is an incredible development tool, with many surprising and powerful advantages. Read on!

Most of what I know about QFD comes from first-hand experience and from comparing notes with other QFD practitioners, who were forced to improvise practical solutions to problems as they encountered them. Indeed, the verbal tradition of QFD in the United States is quite extensive.

The books, articles, and case studies describing QFD and available today arrived too late to help me and my colleagues in 1986. When they did arrive, they focused on what QFD *is* or *could be*. The QFD experience itself has been generally glossed over, as if it were self-evident. Implementing QFD is anything but self-evident! The possibilities for wasting time and leading a team into a cul-de-sac are endless. Many QFD horror stories have at their root the uninformed decisions of an inexperienced QFD facilitator.

On the other hand, the possibilities for helping QFD teams save time and arrive at breakthroughs are abundant. Many QFD success stories have at their root the creative decisions of a capable QFD facilitator.

This book has been written to fill a gap still not addressed by the existing QFD literature in English. This book focuses on the *doing* of QFD. I've tried to address other aspects of QFD as well: parts of this book explain what QFD is, and how QFD can fit in with other organizational activities. But I feel the "QFD Handbook" part of this book is unique in that it provides detailed information on how QFD can successfully be implemented, along with a wide range of choices for customizing QFD, and their advantages and disadvantages.

I've taken special pains to discuss such practical issues as time estimates, group facilitation topics related to QFD, and practical shortcuts. I acquired some of this knowledge the hard way, by making mistakes, and by trying out new ideas and evolving them over the years. I acquired the rest by learning from my colleagues, and by adapting ideas from other disciplines. My goal in this book has been to make the QFD implementation path easier. If the book serves that purpose, then QFD will be used more widely in the United States, and we can hope for better products, better services, and greater market share as a result.

QFD, Software, and Services

Because of my background in software and computers, many of my early QFD experiences were related to planning for new versions of these types of products. For some reason, people seem to think that QFD must be applied differently to software than to other technologies. I have never understood that point of view, except perhaps as an excuse for not using QFD.

The basic problems of product design are universal: customers have needs that relate to using products; the needs must be addressed by designers who have to make hundreds or thousands of technical decisions; and there are never enough people, time, and dollars to put everything that could be imagined into a product or service.

These problems confront the developers of automobiles, cameras, hot-line service centers, school curricula, and even software. QFD can be used to help development teams decide how best to meet customer needs with available resources, regardless of the technology underlying the product or service.

Customers have their own language for expressing their needs. Each development team has *its* own language for expressing its technology and its decisions. The development team must make a translation between the customer's language and their technical language. QFD is a tool that helps teams systematically map out the relationships between the two languages.

Software engineers are fortunate in that *several* languages are available to them for expressing their top-level design: the disciplines of object-oriented design, structured design,[2,3] and structured analysis provide excellent methods for expressing the technical aspects of a software system. These languages can and have been used in software QFDs. Other technical language elements that software engineers have used in QFD are performance measures, subsystem modules, and brief descriptions of product functions. What works best in QFD is the technical language that the development team is most comfortable with. For more on Software QFD, please see Chapter 19.

Examples and Case Studies

In this book, I have freely used examples relating to many technologies and industries. For the most part these examples have come from my own experience, although I have modified them to help clarify the points I was trying to explain in the book. I hope that these examples will convince the reader by demonstration that QFD applies across the board for hardware, software, and service development, as well as for other forms of strategic planning.

I hope the reader will enjoy the Word Processor example that I have used in many places in the book to illustrate various points. I chose this product because I assume most readers will have used word processors, and will identify with the customer needs, if not the internal design. This example is a mixture of pedagogy and fact; I hope it is detailed enough to help you understand QFD, and I am sure that it is sufficiently modified that it will not reveal any trade secrets of any word processor developers.

How to Read This Book

This book is divided into five parts. Each part looks at QFD from a different perspective.

Part I, About QFD, provides motivation for QFD and puts it into perspective within the framework of the global business environment of the last twenty years. It gives the briefest of overviews of what QFD is. It's equivalent to the type of fifteen-minute management summary I often gave to popularize QFD shortly after I first learned about it. It differs from the management overview in that I have tried to compile a chronology of key events in the U.S. that have lead to the popularization of QFD. Read this part if you are new to QFD, or if you would like to find out where it came from and how it has become so well known among product developers.

Part II, QFD at Ground Level, explains QFD in an expository fashion. It's a kind of textbook within a textbook. It explains each portion of the House of Quality in considerable detail. I hope it will be a useful reference for QFD implementors who are looking for detailed information about the HOQ. Read this part if you would like to be fully informed about the House of Quality.

Part III, QFD from 10,000 Feet, assumes the reader has some familiarity with QFD. It provides an organizational perspective on the way product and service development occurs. It shows how QFD can help organizations become more competitive by developing better products and services. Chapter 12 paints a picture of the way the world ought to be, and Chapter 13 discusses the world the way I've experienced it. If you want to introduce QFD into an organization, you'll find the ideas in Part III to be helpful for developing your strategy for organizational change.

Part IV, QFD Handbook, is where the rubber hits the road. It assumes you have decided to implement QFD, and it shows you how to start, what to anticipate, and how to finish successfully. I wish I had had Part IV before I tried my first QFD. Over the years it seems I've made every imaginable mistake, as I've helped teams to use QFD. Part IV assembles the accumulated lessons from my mistakes and from those of many of my colleagues. Read this section before you try to implement QFD. I think it will be a good investment of your time.

Part V, Beyond the House of Quality, points the way to the extensions of the House of Quality that can take the development team all the way to the completion of their project. Although it is used predominantly for product planning, QFD has the potential to help the development team deploy the Voice of the Customer to every phase of development. Part V describes some of the possible paths teams can take after completing the HOQ. It also includes some topics that may be of interest only to some readers. These include specialized adaptations of QFD for software and service development, as well as for organizational planning. Read this part if you are comfortable with QFD concepts and are ready to make QFD the backbone of the entire development process.

Acknowledgments

I am grateful to many people who encouraged me to write this book and who helped me along the way. I am apprehensive about listing their names for fear I have forgotten someone important, but it must surely be better to acknowledge most of my supporters than none at all. First I must acknowledge my good friend and colleague Don Clausing, who started it all in so many ways, both for the nation and for me personally.

As I mentioned earlier, I have learned much of what I know about QFD from my fellow QFD practitioners. In addition, two organizations, the American Supplier Institute (ASI) and GOAL/QPC, have done much to train U.S. industry in QFD. For various reasons, geography among then, my personal contact with GOAL/QPC has been much more direct and extensive than with ASI. ASI's influence on my understanding of QFD is indirect, but still quite substantial, since so many of my QFD colleagues have been associated with ASI. I thank both these organizations, and their able leaders, Larry Sullivan and Bob King, for helping establish QFD as a key product development tool in the U.S.

For showing their faith in me and my message about QFD, my eternal thanks go to Russ Doane, Bob King, Cristina Davy, Catherine D'Abadie, Susan Ellis-Richard, Abbie Griffin, John Hauser, Kurt Hoffmeister, Karen Holtzblatt, Tony Hutchings, Michael Hyde, Gerry Katz, Takami Kihara, Bob Klein, Mel Klein, Ramon Leon, Mary Ellen Lewandowski, Glenn Mazur, Jim Mills, Leslie Sisto, Rachel Oberai, Yogesh Parikh, Madhav Phadke, Bill Pardee, Ed Prentice, John Terninko, Steve Ungvari, John W. Wesner, and Richard E. Zultner.

Special thanks to my appropriately impatient and eternally helpful editor, Jennifer Joss, and to many others at Addison-Wesley whose professionalism has been such a great benefit to this book, including Tara Herries, Marty Rabinowitz, and Chris Sykes.

---

[1] Credit is generally given to Harold Ross (General Motors) and Bill Eureka (American Supplier Institute) for the "What/How" terminology, which is now widely used by QFD practitioners in the U.S.

[2] Edward Yourdon and Larry L. Constantine, "Structured Design: Fundamentals of a Discipline of Computer Program and System Design," Yourdon Press, A Prentice-Hall Company, 1979.

[3] Chris Gane and Trish Sarson, "Structured Systems Analysis: Tools and Techniques," MCAUTO McDonnell Douglas, 1982.

PART I

About QFD

CHAPTER 1

What Is QFD?

Introduction

This chapter provides an orientation to QFD. It starts with a brief "management" overview. It then provides a chronology of events that led to the current state of QFD in the United States. Finally, the chapter discusses some of the ways QFD is being used today, to provide a sense of the applicability and flexibility of the process.

1.1   Brief Capsule Description

QFD (Quality Function Deployment) is a method for structured product planning and development that enables a development team to specify clearly the customer's wants and needs, and then to evaluate each proposed product or service capability systematically in terms of its impact on meeting those needs.

The QFD process involves constructing one or more matrices (sometimes called "quality tables"). The first of these matrices is called the "House of Quality" (HOQ). It displays the customer's wants and needs (the "Voice of the Customer") along the left, and the development team's technical response to meeting those wants and needs along the top. The matrix consists of several sections or submatrices joined together in various ways, each containing information related to the others (see Diagram 1-1).

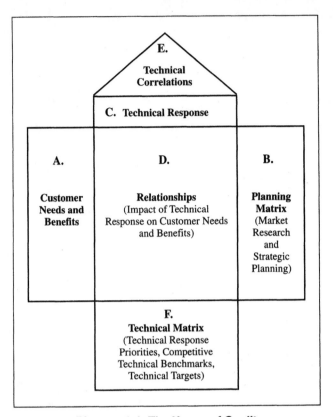

Diagram 1-1. The House of Quality

Each of the labeled sections, A through F, is a structured, systematic expression of a product or process development team's understanding of an aspect of the overall planning process for a new product, service, or process. The lettering sequence suggests one logical sequence for filling in the matrix.

Section A contains a structured list of customer wants and needs. The structure is usually determined by qualitative market research. The data is in the form of a tree diagram (defined and explained in Chapter 3).

Section B contains three main types of information:

- Quantitative market data, indicating the relative importance of the wants and needs to the customer, and the customer's satisfaction levels with the organization's and its competition's current offerings
- Strategic goal setting for the new product or service
- Computations for rank ordering the customer wants and needs

Section C contains, in the organization's technical language, a high-level description of the product or service they plan to develop. Normally this technical description is generated (deployed) from the customer's wants and needs in Section A.

Section D contains the development team's judgments of the strength of the relationship between each element of their technical response and each customer want and need.

Section E, Technical Correlations, is half of a square matrix, split along its diagonal and rotated 45°. Since it resembles the roof of a house, the term "House of Quality" (HOQ) has been applied to the entire matrix and has become the standard designation for the matrix structure. Section E contains the development team's assessments of the implementation interrelationships between elements of the technical response.

QFD authorities differ in the terminology associated with parts of the House of Quality. I have tried to be consistent throughout this book, but in day-to-day usage most people use various terms interchangeably. I have tried to indicate alternate terminology whenever I have introduced a new term. There is no reason to be concerned about the lack of standardization; it rarely causes actual confusion.

Section F contains three types of information:
- The computed rank ordering of the technical responses, based on the rank ordering of customer wants and needs from Section B and the relationships in Section D
- Comparative information on the competition's technical performance
- Technical performance targets

Beyond the House of Quality, QFD optionally involves constructing additional matrices which further guide the detailed decisions that must be made throughout the product or service development process. In practice many development teams don't use matrices after the House of Quality. They are missing a lot. The benefits that the House of Quality provides can be just as significant to the development process after the initial planning phase. I urge you to use Part V of this book to become familiar with the later stages of QFD, and then, with that knowledge available to you, make an informed decision about how much of QFD your development project needs.

Diagram 1-2 illustrates one possible configuration of a collection of interrelated matrices. It also illustrates a standard QFD technique for carrying information from one matrix into another. In Diagram 1-2 we start with the House of Quality (HOQ). We place the "Whats" on the left of the matrix. "Whats" is a term often used to denote benefits or objectives we want to achieve. Most

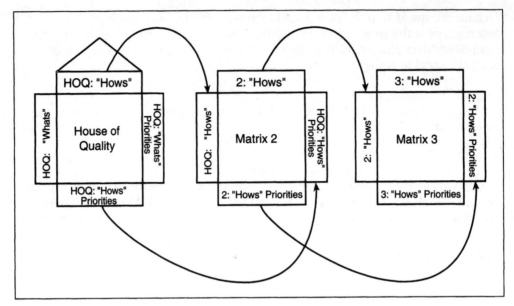

Diagram 1-2. Interrelated Matrices

commonly, the "Whats" are the customer needs, or the Voice of the Customer, but the development team's own objectives could also be represented as "Whats." As part of the QFD process, the development team prioritizes the "Whats" by making a series of judgments based in part on market research data. There are many different techniques for determining these priorities, described later in this book. The priorities or weights are placed to the right of the matrix.

Next, the development team generates the "Hows" and places them along the top of the matrix. The "Hows" are any set of potential responses aimed at achieving the "Whats." Most commonly, the "Hows" are technical measures of performance of the proposed product or service.

Based on the weights assigned to the "Whats" and the amount of impact each "How" has on achieving each "What," the "Hows" are given priorities or weights, which are written at the bottom of the HOQ. These weights are a principal result of the HOQ process.

To link the House of Quality to Matrix 2, the development team places all, or the most important, of the HOQ "Hows" on the left of Matrix 2, and their priorities on the right of Matrix 2. These HOQ "Hows" now become the "Whats" of Matrix 2, and their relative importance to the development team is what was determined in the HOQ.

To achieve the Matrix 2 "Whats," the development team needs a new, more technical or more detailed set of "Hows," which they generate and put at the top of Matrix 2. As before, the team uses the weights of the Matrix 2 "Whats," and their estimates of the degree of relationship between the Matrix 2 "Hows" and the Matrix 2 "Whats," to arrive at weights or priorities for the Matrix 2 "Hows."

To link Matrix 2 to Matrix 3, the Matrix 2 "Hows" are transferred to the left of Matrix 3, to become the Matrix 3 "Whats." The Matrix 2 "Hows" weights are transferred to the right of Matrix 3, and new Matrix 3 "Hows" are generated.

Each matrix in the chain represents a more specific or more technical expression of the product or service. In the classical model for QFD,[1] which was based on the development of hardware products, the relationship of "Whats" to "Hows" in each matrix is shown in Diagram 1-3.

This model mirrors the process of designing and manufacturing a product. Similar models exist for developing services, for designing processes, and for developing software products. They are all covered in other parts of this book.

Other multiple matrix QFD schemes are considerably more elaborate than the four-matrix scheme described in Diagram 1-3.[2] Some QFD matrix schemes involve as many as thirty matrices that use the VOC priorities to plan multiple levels of design detail, Quality Improvement Plans, Process Planning, Manufacturing Equipment Planning, and various Value Engineering plans.

Some QFD experts believe that the use of additional matrices is not optional, and that the process should not be called QFD unless a series of matrices are constructed. In this book, however, we take a more liberal view. We take the position that QFD delivers many possible benefits. Depending on the benefits a development team needs or is willing to work for, they will construct just the initial House of Quality, or a large collection of interrelated matrices, or something in between. They will further customize their matrices to solve the problem they need to solve. From our point of view, it's all QFD. Applying the old adage "Science is what scientists do," we believe that "QFD is what QFD practitioners do."

Matrix	What	How
House of Quality	Voice of the Customer	Technical Performance Measures
Subsystem Design Matrix	Technical Performance Measures	Piece-Part Characteristics
Piece Part Design Matrix	Piece-Part Characteristics	Process Parameters
Process Design Matrix	Process Parameters	Production Operations

Diagram 1-3. Classical Model for QFD

As we have seen in Diagram 1-3, QFD is a tool that enables us to develop project priorities at various levels in the development process, given a set of priorities at the highest level (customer needs). QFD is not just a prioritization tool, it's also a "deployment tool." What we mean by "deployment" is that QFD helps us to start with the highest level of "Whats," generally the Voice of the Customer, and to deploy, or translate, that voice into a new language that opens the way for appropriate action. Every developer is familiar with this translation process and probably performs the translation informally all the time. QFD helps make the translation process explicit and systematic.

Finally, QFD provides a repository for product planning information. The repository is based on the structure of the QFD matrices. The matrices allow for entering the Voice of the Customer (and, in subsequent matrices, other deployed "What" and "How" information) and all related quantitative information, the Voice of the Developer and all related quantitative information, and the relationships between these voices. This information represents a succinct summary of the key product planning data. To be sure, much detailed elaboration of this information cannot fit in the QFD matrices and must be stored elsewhere. However, the matrices can be viewed as a top-level view or directory to all the rest of the information. In fact, QFD has sometimes been described as a "visible memory of the corporation."[3]

There are many other views of QFD, and they will be explored elsewhere in this book. For now, it will be useful to keep in mind these main ideas: QFD provides a formal linkage between objectives ("What") and response ("How"). It also assists developers in developing or deploying the "Hows" from the "Whats," it provides a systematic method of setting priorities, and it provides a convenient repository of the information.

1.2   History of QFD

The people I've identified in this chronology have made special efforts in the best interests of U.S. industry. They could have kept QFD as a proprietary secret, not to be shared with the competition. Instead, they shared their experiences with others, including their competitors, and everyone has gained.

QFD became widely known in the United States through the efforts of Don Clausing, of Xerox and later MIT, and Bob King of GOAL/QPC. These two worked independently from each other and likely first came in contact in October 1985, when Clausing presented QFD at a GOAL/QPC conference in Massachusetts. By that time, both men had already made significant contributions toward promoting QFD.

The Japanese characters for QFD are, phonetically,

- Hinshitsu, meaning "quality," "features," "attributes," or "qualities"
- Kino, meaning "function" or "mechanization"
- Tenkai, meaning "deployment," "diffusion," "development," or "evolution"

Any of the English words could have been chosen by early translators of Japanese articles. It's little more than a matter of chance that QFD is not called Feature Mechanization Diffusion today. In the early days, when the author explained QFD to audiences, he attempted to rename it as Structured Planning, or Quality Feature Deployment, in the hopes that people would be able to tell from its name what QFD was all about. For better or for worse, "Quality Function Deployment" has stuck in the United States, and no alternative name is likely to survive. None of the thirty-two possible combinations of English equivalents really denotes what QFD actually is. We must be content with a name for the process which is not self-explanatory.

Don Clausing first learned about QFD in March 1984, during a two-week trip to Fuji-Xerox Corporation, a Xerox partner in Japan. Clausing, a Xerox employee at that time, had already become interested in the Robust Design methods of Dr. Genichi Taguchi, who was a consultant to Fuji-Xerox. While in Japan, Clausing met another consultant to Fuji-Xerox, a Dr. Makabe of the Tokyo Institute of Technology.

At an evening meeting, Dr. Makabe briefly showed Clausing a number of his papers on product reliability. "After fifteen minutes," relates Clausing, "Dr. Makabe brushed the papers aside and said, 'Now let me show you something *really* important!' Makabe then explained QFD to me. I saw it as a fundamental tool that could provide cohesion and communication across functions during product development, and I became very excited about it."

In the summer of that same year, Larry Sullivan of Ford Motor Company organized an internal company seminar. Clausing was invited to present QFD. Sullivan quickly grasped the importance of the QFD concept and began promoting it at Ford.

Clausing continued to promote QFD, Taguchi's methods, and Stuart Pugh's concept selection process at conferences and seminars. When he joined the faculty at MIT, he developed a semester-length graduate course that unified these methods along with other concepts into a system for product development that eventually became called "Total Quality Development." Many of his students, already senior managers and engineers at large U.S. companies, returned to their jobs and spread his concepts to their co-workers.

In June 1987, Bernie Avishai, associate editor of *Harvard Business Review,*
asked Don Clausing to write an article on QFD. Clausing felt that the paper
should be given a marketing perspective, and he invited John Hauser to co-
author it. Hauser had become intrigued with QFD after learning about it from
a visit to Ford. The article, published in the May–June 1988 issue of the *Harvard Business Review,* has become one of the publication's most frequently
requested reprints. That article probably increased QFD's popularity in the
United States more than any other single publication or event.

Larry Sullivan founded the Ford Supplier Institute. This was a Ford Motor
Car organization aimed at helping Ford's suppliers improve the quality of the
components they developed for Ford. Sullivan and others at Ford gained a
detailed understanding of QFD by working with Dr. Shigeru Mizuno and Mr.
Akashi Fukahara from Japan. Eventually Ford came to require its suppliers to
use QFD as part of their development process, and the Ford Supplier Institute
provided QFD training (along with other topics) to these suppliers.

The Ford Supplier Institute eventually became an independent nonprofit
organization, called the American Supplier Institute (ASI). ASI has become a
major training and consulting organization for QFD. It has trained thousands
of people in the subject.

Bob King, founder and executive director of GOAL/QPC, first learned of QFD
from Henry Klein of Black and Decker. Klein had attended a presentation on
QFD given by Yoji Akao and others in Chicago, in November 1983. The following month, Klein attended a GOAL/QPC course on another TQM topic,
where he told King about this presentation and about QFD. King began offering courses on QFD, starting in March 1984. In the summer of that year, King
learned more details about QFD from a copy of a 1978 book by Akao and
Mizuno, called *Facilitating and Training in Quality Function Deployment.* In the
fall of 1984, King began offering a three-day course on QFD, based on the
understanding of the tool he had gained from the Akao and Mizuno book. In
November 1985, King traveled to Japan and met with Akao to "ask him all the
questions he couldn't answer." Akao provided King with his course notes on
QFD, and he gave GOAL/QPC permission to translate the notes and use them
in his GOAL/QPC courses.

Based on these notes, GOAL/QPC offered its first five-day course on QFD in
February 1986. This author attended that course and learned about QFD there
for the first time.

At the invitation of Bob King, Akao came to Massachusetts and conducted a
workshop on QFD in Japanese, with simultaneous translation into English.
Akao conducted a second workshop in June 1986, also under the auspices of
GOAL/QPC. For this second workshop, GOAL/QPC translated a series of

papers on QFD, including several case studies. This translation was later published in book form.[4] Eventually this collection of QFD papers became what remains the standard advanced book on QFD.

In 1987, GOAL/QPC published the first full-length book on QFD in the United States: *Better Designs in Half the Time*, by Bob King.[5] In this book, King described QFD as a "matrix of matrices" (see Chapter 18). King relates that in June 1990, Cha Nakui, a student of Akao's and later an employee in Akao's consulting company, "comes to work for GOAL/QPC and corrects flow of QFD charts." Among Nakui's contributions to our understanding of QFD is his explanation of the Voice of the Customer Table (see Chapter 5).

Other important early publications in the United States include

- "Quality Function Deployment and CWQC in Japan," by Professors Masao Kogure and Yoji Akao, Tamagawa University, published in *Quality Progress* magazine, October 1983
- "Quality Function Deployment," by Larry Sullivan, published in *Quality Progress* magazine, June 1986
- Articles on QFD by Bob King and Lou Cohen in the spring and summer 1988 editions of the *National Productivity Review*
- A series of articles on QFD in the June 1988 issue of *Quality Progress* magazine
- A course manual on QFD to supplement ASI's three-day QFD course
- *Annual Proceedings* of QFD symposia held in Novi, Michigan, starting in 1989

QFD software packages first became available in the United States around 1989. The most widely known package, "QFD/Capture," was developed by International TechneGroup Incorporated. Some organizations, heavily committed to QFD, such as Ford Motor Company, have developed their own QFD software packages.

Early adapters of QFD in the United States included Ford Motor Company, Digital Equipment Corporation, Procter and Gamble, and 3M Corporation. Many other companies have used QFD, and the tool continues to grow in popularity. More than fifty papers were presented at the Sixth Symposium on Quality Function Deployment in 1994. Only a few of these papers were not case studies. The majority of companies using QFD are reluctant to present their case studies publicly, since they don't want to reveal their strategic product planning. Therefore, it is likely that the fifty papers presented at the QFD Symposium represent just the tip of the iceberg in terms of QFD implementation.

Diagram 1-4 identifies many key events in the development of QFD, both in Japan and the United States. In some cases, where exact dates are not known, approximate months or years have been provided.

Date	Source	Event
1966	*Facilitating and Training in Quality Function Deployment,* Marsh, Moran, Nakui, Hoffherr	Japanese industry begins to formalize QFD concepts developed by Yoji Akao
1966	*Facilitating and Training in Quality Function Deployment*	Bridgestone's Kurume factory introduces the listing of processing assurance items: "Quality Characteristics"
1969	*Facilitating and Training in Quality Function Deployment*	Katsuyoshi Ishihara introduces QFD at Matsushita
1972	*Facilitating and Training in Quality Function Deployment*	Yoji Akao introduces QFD quality tables at Kobe Shipyards
1978	*Facilitating and Training in Quality Function Deployment*	Dr. Shigeru Mizuno and Dr. Yoji Akao publish *Deployment of the Quality Function* (Japanese book on QFD)
1980	*Facilitating and Training in Quality Function Deployment*	Kayaba wins Deming prize with special recognition for using Furukawa's QFD approach for bottleneck engineering
1983	*Facilitating and Training in Quality Function Deployment*	Cambridge Corporation of Tokyo, under Masaaki Imai, introduces QFD in Chicago along with Akao, Furukawa, and Kogure
10/83	*Quality Progress* magazine	"Quality Function Deployment and CWQC in Japan," by Professors Masao Kogure and Yoji Akao, Tamagawa University
11/83	Bob King	Akao and others introduce QFD at a U.S. workshop in Chicago, Illinois
3/84	Don Clausing	Professor Makabe of Tokyo Institute of Technology explains QFD to Don Clausing
3/84	Bob King	Bob King begins offering a one-day course on QFD
7/17/84	Don Clausing	Don Clausing presents QFD to a Ford internal seminar organized by Larry Sullivan
1985	*Facilitating and Training in Quality Function Deployment*	Larry Sullivan and John McHugh set up a QFD project involving Ford Body and Assembly and its suppliers
10/30/85	Don Clausing	Don Clausing presents QFD at GOAL/QPC's annual conference
11/85	Bob King	King meets with Akao in Japan. Akao gives GOAL/QPC permission to translate his classroom notes and use them in GOAL/QPC's training
1/27/86	Don Clausing	Don Clausing presents QFD to Ford's Quality Strategy Committee #2, chaired by Bill Scollard of Ford
2/86	*Facilitating and Training in Quality Function Deployment*	GOAL/QPC introduces Akao's materials in its five-day QFD course (the author attended this course in February 1986)
6/86	Don Clausing	Larry Sullivan sponsors Dr. Mizuno who gives a three-day seminar on QFD

6/86	*Quality Progress* magazine	"Quality Function Deployment" by Larry Sullivan
10/86	Don Clausing	Larry Sullivan launches QFD at Ford
6/87	Don Clausing	Bernie Avishai, associate editor of *Harvard Business Review*, asks Don Clausing to write an article on QFD. Don invites John Hauser to co-author it. It is published in May–June 1988
1989–present	ASI, GOAL/QPC	Sponsorship of QFD Symposia at Novi, Michigan
2/6/91	Don Clausing	Don Clausing and Stuart Pugh present "Enhanced Quality Function Deployment" at the Design and Productivity International Conference, Honolulu, Hawaii

Diagram 1-4. History of QFD in the United States

1.3   What Is QFD Being Used for Today?

As with any versatile tool, the applications of QFD are limited only by one's imagination. The original intent of QFD was to provide product developers with a systematic method for "deploying" the Voice of the Customer into product design. The need to evaluate potential responses against needs is universal, however, and in the United States a wide range of applications sprang up quite rapidly.

Here are some typical QFD applications that do not fit the model of product development:

- Course design: "Whats" = needs of students for acquiring skills or knowledge in a certain area; "Hows" = course modules, course teaching style elements. Curriculum design is a natural extension of this application and has also been done using QFD[6]
- Internal corporate service group strategy: "Whats" = business needs of individual members of service group's client groups; "Hows" = elements of the service group's initiatives
- Business group five-year product strategy: "Whats" = generic needs of the business group's customers; "Hows" = product offerings planned for the next five years
- Development of an improved telephone response service for an electric utility company[7]: "Whats" = the needs of the customer; "Hows" = critical measures of performance of the telephone answering center

In the author's experience, those who become enthusiastic about QFD are generally very creative in conceiving new applications. Those who dislike the "matrix" approach to planning are generally very creative in producing reasons why QFD doesn't work.

Of course, QFD itself does not "work." Just as we cannot say that hammers "don't work," so also we cannot say that problem-solving tools such as QFD "don't work." A more accurate statement would be: "I can't work it." This book is intended to raise your confidence that, when it comes to QFD, you can say, "I can make it work for me."

Discussion Questions

Who is your customer? Try writing down your answer.

Think about your development process today, for developing products, services, courses, strategies, or anything else. Do you know your customers' needs? How do you know them? What form do you use to represent those needs?

What does your first formulation of "Hows" look like? What form does it take? How do you generate your "Hows"?

How do you determine the relationship of your "Hows" with your customers' needs?

How many levels of "Hows" do you have in your development process? Write down a brief description of each of these levels and show them to your colleagues. Notice how long it takes to get agreement.

Besides your development process, where else in your daily work might QFD apply? Define the "Whats" and "Hows."

[1] *Total Quality Development*, by Dr. Don Clausing, ASME Press, 1994.

[2] *Better Designs in Half the Time*, by Bob King, GOAL/QPC, Methuen, Massachusetts, 1987.

[3] Quote attributed to Max Jurosek, Ford Motor Company, from "Enhanced Quality Function Deployment," series of five instructional videotapes, MIT Center for Advanced Engineering Study (no copyright date appears on these tapes).

[4] *Quality Function Deployment: Integrating Customer Requirements into Product Design*, Yoji Akao, author and editor-in-chief, Tamagawa University, Japan, translated by Glenn H. Mazur and Japan Business Consultants, Ltd., Productivity Press, Cambridge, Massachusetts, 1990.

[5] *Better Designs In Half the Time*, by Bob King, GOAL/QPC, 1987.

[6] Mahesh Krishnan and Dr. Ali A. Houshmand, "QFD In Academia: Addressing Customer Requirements in the Design of Engineering Curricula," presented at the Fifth Symposium on Quality Function Deployment, June 1993, Novi, Michigan.

[7] Amy Tessler, Norm Wada, and Bob Klein, "QFD at PG&E—Applying Quality Function Deployment to the Residential Services of Pacific Gas & Electric Company," presented at the Fifth Symposium on Quality Function Deployment, June, 1993, Novi, Michigan.

CHAPTER 2

How QFD Fits in the Organization

Introduction

This chapter sets a context for QFD in terms of the problems QFD helps us solve. The fundamental business strategies for competitiveness—decreasing costs, increasing revenues and reducing the time to produce new products and services—can all be enhanced by QFD. This chapter explains QFD's role with respect to each. One of the most important product development process concepts in recent years, specifically aimed at reducing product development time, is Concurrent Engineering (CE). This chapter will introduce CE and show how QFD is critically connected to it.

Finally, we'll explore the Kano model for customer satisfaction. The model expands our traditional views of how customers are satisfied by products and services. It provides some important food for thought as we explore the way in which QFD uses the Voice of the Customer.

2.1   The Challenge to the Organization

An organization's motivations are very similar to an individual's motivations. This analogy between individuals and organizations is quite natural, since organizations are made up of individuals.

As individuals, we first need food and shelter. Once we have these, we seek steady supplies of food and shelter—in other words, security. If our personal security can be taken for granted, we might turn our attention to more abstract goals, such as expanded influence on the people around us. However, if our source of food and shelter is threatened, we are likely to forsake such goals and concentrate again on establishing reliable supplies of food and shelter—that is, on regaining security.

Diagram 2-1 shows this schematically: Increased ability to survive leads to increased security. Once an individual or an organization is secure, the possibility of expansion presents itself. As the organization expands, survival becomes more certain.

The organizational analogue is this: food and shelter for the individual correspond to enough money flowing into a corporation to allow it to pay its bills and for everyone already there to draw a salary. (The salaries provide the food and shelter required by each individual.) Steady supplies of food and shelter are equivalent to relatively constant flows of revenue, accompanied by adequate margins, so that the organization has positive cash flow. An organization's more abstract goals, beyond mere survival, may include growth and increased influence upon its environment.

Just as an individual's food and shelter supplies may be threatened, so can an organization's revenue flow, profit margin, and cash flow be threatened.

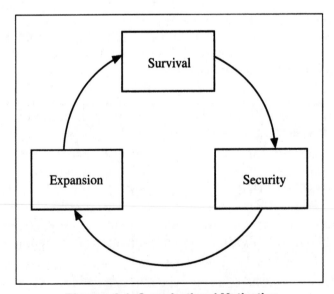

Diagram 2-1. Organizational Motivation

Obviously the most common threat to revenue flow is the competitive threat—another organization offering products or services that meet the same needs as our organization is meeting. Other threats arise from rising costs or waste within the organization.

In this never-ending cycle to deal with threats to survival, security, and expansion, organizations are continually evolving. The organization most successful in coping with these threats has the greatest ability to expand its influence, and therefore imposes threats on other similar organizations with similar goals.

QFD can play an important role in helping organizations become stronger, and therefore more likely to survive, more secure, and more able to expand. In this chapter we'll explore various ways that QFD can help. A useful way of thinking about this is to divide an organization's coping strategies into two main categories: decreasing its costs, and increasing its revenues.

2.2   Decreasing Costs

Decreased costs can be achieved by such actions as lowering the cost of purchased materials or services, reducing overhead costs for the office or plant, or reducing payroll.

Decreased costs can also be achieved by streamlining processes and reducing rework and waste. QFD contributes to decreased costs by this latter approach. It does so by the following means:

- Increasing the likelihood that a product or process design will not have to be changed or redone. This "dampening" effect comes about because QFD allows developers to evaluate proposed midproject changes against the same criteria used to evaluate all design decisions at the beginning of the project. The team has simply to add the new proposed change to their QFD matrices and apply the same analysis to it that they applied to all the earlier decisions. This systematic analysis helps developers avoid panicky, rushed decisions that fail to take the entire product and all the customer needs into account. Most "midcourse" corrections are easily rejected or postponed when QFD analysis is applied to them.
- Focusing product and process development on the work that matters the most to the customer, rather than on work that means little or nothing to the customer. This is another way of saying the work that gets done is what QFD analysis has shown to be most clearly related to meeting customer needs.

2.3   Increasing Revenues

Increased revenues are normally achieved by selling more of a product or service, or by charging more for the product or service. Both of these desirable results can be attained by producing products or services that are more attractive to customers. Products or services are made more attractive to customers by better meeting their needs.

QFD contributes to increased revenues by helping organizations to concentrate their efforts on customer needs, and to accurately and effectively translate customer needs into the right product design or the right service characteristics.

2.4   Cycle Time Reduction

Most organizations have competition. Each price change, each product announcement, each reorganization—in fact, each change an organization makes—is like a move in a game between competitors. Each move affects an organization's competitiveness, for better or for worse. Often the results cannot be observed immediately, nor can an organization regain lost competitiveness rapidly. Thus, the stakes of the game are high.

One of the most important "moves" is to make new products or services available before the competition does so. By shipping a new product before a competitor can offer an equivalent one, the competition can be robbed of business that may be impossible to regain. Many believe today that rapid development of new products or services is the single most important key to competitiveness—in other words, it's the winning "move."

Since most organizations begin designing new versions of products or services as soon as the previous ones have been released, the product development process is usually viewed as a "cycle." In fact, the development process, along with all other processes, is not quite so neat as a simple circle, as in Diagram 2-2. Often work on the *next* product begins well before the *previous* product is ready to be sold. Many other simultaneous activities are usually happening as well. Nevertheless, the "cycle" model is a helpful way of viewing the activity.

QFD is an important key to cycle time reduction. Throughout this book, we'll be seeing how QFD helps development teams make key decisions early in the development process, at a time when the cost of a decision is relatively low.

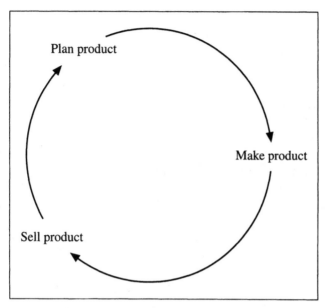

Diagram 2-2. Product Development Cycle

Some of the ways that QFD contributes to reduced cycle time are the following:

- QFD helps reduce midcourse changes. Midcourse changes, such as shifts in priorities, key vendor replacements, or replacement of key technologies, wreak havoc on development schedules. When a project is well underway and a major change in direction is proposed, most development teams are unable to evaluate the proposal properly. This is because there is enormous schedule pressure on teams, and that pressure normally increases as development proceeds. Consequently, midcourse changes are often made rapidly, without due consideration to their overall impact.

  When a project has been planned using QFD, the House of Quality and other matrices provide a summary-at-a-glance of the project strategy. A well-done QFD planning process will clearly lay out all the needs of the customer, and will show how the strategic and design decisions of the project relate to those needs. Midcourse proposals can rapidly be added to the QFD matrices and evaluated in the context of all the previously made decisions.

  Very few midcourse changes actually survive this type of thorough analysis.

- QFD helps reduce errors in implementation. By working through the QFD process, the development team ensures a common vision of the customer needs and their agreed-upon responses to them. This common vision, along with detailed planning all starting from the same place (the House of Quality), results in consistent follow-through across the entire development team.

2.5   Obstacles to Rapid Product Development

A key to competitiveness is the ability to respond to the competition by pro-ducing new products and services rapidly. There are many obstacles to rapid production of products and services. Some of these are

- Poor understanding of customer needs
- Failure to strategically prioritize efforts
- Willingness to take on unmanageable risks
- Tendency toward unbuildable designs, undeliverable services
- Overreliance on formal specifications
- Testing scenarios that fail to find key defects

2.5.1   Poor Understanding of Customer Needs

Inadequate research into customer needs leads to varied, uninformed opin-ions. This leads to disagreement among the developers as to customer needs. Such disagreements in turn create delays in decision-making, reversals of decisions, and individuals working at cross-purposes.

QFD provides a standardized method of representing customer needs. This method is by no means a method of *learning* what the customer needs are (acquiring the Voice of the Customer is discussed in other parts of this book), but it does provide a way of systematically representing those needs. The standard representation can then be used as a basis for itemizing differences of opinion, which can be researched as needed. (The best approach, of course, is to perform credible market research up front to eliminate the chance for these differences of opinion. However, not all developers do so.)

QFD's standardized method of *mapping* these customized needs to product or service development decisions helps further reduce differences of opinion.

2.5.2   Failure to Strategically Prioritize Efforts

One key to cycle-time reduction is to invest in what's important and to resist investing in what's not important.

Failure to take time to work on capabilities that are important to customers leads to noncompetitive products. Such failures, in turn, often lead to longer cycle times, because the overlooked capabilities must be added after the product or service has been designed. Generally this type of midcourse correction is very error-prone, because it is done hastily, and because it is attempted as an overlay to an existing design.

Taking time to work on product capabilities that matter little to customers can only increase cycle time with no benefit to the customer or to revenues. Often these unnecessary efforts add cost to products or services and delay their introduction, which makes them less competitive.

QFD helps developers decide on the relative importance of their choices by deriving *their* priorities from their *customers'* priorities.

2.5.3   Willingness to Take on Unmanageable Risks

All too often, product developers plunge into projects that have unnecessary risks built in. Typically these risks relate to unmanufacturable and unserviceable designs. Often these risks are assumed in desperation. Developers feel forced to commit to schedules they can't meet. Manufacturing managers and engineers feel forced to commit to volumes, quality levels, and cost limitations they can't meet. QFD helps make these risks more visible and easier for development teams to plan for up front. Also, Pugh's concept selection process, an important adjunct to QFD, helps teams to identify or synthesize less risky alternatives.

2.5.4   Tendency toward Unbuildable Designs, Undeliverable Services

Typically, products and services are conceived of by a core team. After they have worked out the system-level details to a certain point, they will normally decompose the design into subsystems or subservices. They will decide that certain subsystems or subservices will be purchased from external or internal suppliers.

The normal method of requisitioning is based on conveying a specification for the desired subsystem or subservice to the supplier organization, which is held at arm's length. The supplier works on an assigned task, and when finished, or at designated intermediate points, presents its results to the core team.

This "throw it over the wall" method often leads to unsatisfactory, even disastrous results, because the supplier's subsystem or subservice, while meeting the specification, will not have completely met the ultimate customer's needs.

The underlying problem lies in the myth that a well-written specification is all that is needed to ensure the desired results. This is our next obstacle:

2.5.5   Overreliance on Formal Specifications

The specification by itself, whether verbal or written, whether a page of text or a thousand pages, can never express all that is required. Deming taught us this years ago.[1] We often operate under the myth that we should be able to write down everything that is needed in a subsystem. But this is rarely the case.

Developing a subsystem by simply meeting a specification is a lot like driving a car without looking at the road (Diagram 2-3). Imagine simply turning left or right when told to by a navigator sitting next to the driver. There are too many details about the road, the speed of the car, the positions of other cars, people and objects in the road, which must be taken into account when a driver makes a turn. These things can't be communicated accurately or in time for a safe journey.[2]

Don Clausing has pointed out that product and service developers make many decisions every day, perhaps thousands of decisions over the course of the project. To be sure, most of these decisions are detailed in scope, but the sheer quantity of them dramatically affects the final result. Even if the answer to every decision were to be found in some voluminous specification, we cannot imagine that a developer will take the time to consult the specification for every one of thousands of decisions.

In order to ensure that each developer, both in the core team and in the supplier's teams, is making daily decisions that all come together in a unified whole that meets customer needs, some other approach to communicating requirements is necessary.

Diagram 2-3. Specifying How to Drive

QFD helps in this area by providing a focus for discussion and for translation of customer needs to each level of the developer and supplier organization, using terminology and language appropriate to that level. The discussion is as important as the resulting specification, because it carries with it all the subtle nuances of meaning that qualify and elaborate on the static language of the specification. In a sense, the discussions inherent in QFD make the specification come alive.

2.5.6   Testing Scenarios That Fail to Find Key Defects

Testing programs find plenty of defects in products and services. Ask any developer who is trying to meet tight deadlines and who must submit work to a tester before it can be called complete! In complex products there are usually so many defects found in internal testing that the developers spend a great deal of time debating which ones to fix and which ones to ship with the product.

Despite all the internal testing, all too often we see customers complain about the defects they have just purchased. Given the embarrassment[3] of defects awaiting the unsuspecting customer, what's needed is a proactive strategy for testing and defect-fixing that concentrates on eliminating those defects most important to the customer—in other words, prioritizing the defects.

Despite Deming's remonstration to cease dependence on mass inspection, abundant defects exist in complex designs and complex services. Therefore, design defect management remains an unhappy fact of life.

QFD provides a method for linking customer needs, and their relative importance, to all development activities, including testing and defect repair. QFD can be used to prioritize testing and repair activities so as to best meet customer needs.

2.5.7   QFD: An Obstacle to Rapid Development?

QFD is time-consuming. Worse than that, it is *explicitly* time-consuming, in the sense that QFD makes obvious and visible the need for several long meetings, attended by quite a few people. For groups that have never used QFD before, this appears as time added to their already crowded schedules. What's not as explicit or visible is the time that QFD saves.

The first step in any development activity is a "Requirements Phase" in which key decisions are made that determine the path of the rest of the development process. In most development environments, this process is experienced by the development team as unstructured, endlessly recycling, mysterious, and unsatisfactory. The QFD process systematically guides a development group through a process of answering questions and making judgments that are

exactly those they should be dealing with in order to determine the requirements. The difference between QFD and less structured requirements-setting
processes is that QFD forces development teams to consider *all* the issues,
especially all the customer needs, in a systematic fashion.

It might seem that a comprehensive look at all issues could take longer than
most development groups customarily spend. However, experience has
shown that the QFD process generally covers more ground faster than less
structured methods. Not only can QFD reduce the overall development
process by providing more complete planning, it also takes less time than less-
structured planning methods.

This is because QFD provides a *process*, a clear set of steps, for making the up-
front decisions that constitute the requirements phase of development. QFD
can be seen as a planning road map that informs the development team of
what decisions must be made at each step, and what information is needed to
make those decisions. This road map helps the team to plan in such a way that
they will have the information they need when it's time to make decisions,
thus reducing the likelihood that they will need to revisit previous decisions.

2.6   QFD's Role as Communication Tool

In the previous section we reviewed many of the obstacles to rapid development of products and services. Obviously many factors could explain why these
obstacles occur at all. One factor, poor communication, emerges frequently as
the enemy of all efficient group processes. It is so much a common theme that
it must be regarded as a major cause of delays in product development.

QFD provides a method for individuals involved at various steps of the development process to communicate with each other. It does this by effectively
translating the language of one phase of development into the language of the
next. The fact that representatives from each phase perform these translations
together further increases the likelihood that the translation will be understood by everyone.

Once the development team has completed its translation and recorded it in a
QFD matrix, the matrix can be presented and explained to others who were
not present during the QFD process. These people can examine the translation
process in minute detail if they need to, focusing on areas that are of special
interest to them and skipping over other areas.

QFD's contribution to improving communication is then threefold:
  • Standard format for translating "Whats" to "Hows"
  • Helps people focus on facts, rather than feelings

- The decision process is recorded in the matrices and can be reexamined and even modified at any time

2.7   Concurrent Engineering: A Paradigm Shift

Concurrent Engineering is the ideal of planning and implementing all product development steps, from early product conceptualization to delivery and service, as early as possible. Team members are responsible for each step, working together throughout.[4] Consider the schematic of a typical product development process that does *not* employ concurrent engineering (Diagram 2-4).

In the traditional product development process, each step is conceived of as a unit with clear inputs and outputs. Steps further downstream, such as Manufacturing Process Development, are not supposed to start until the results of previous steps, such as Component Design, are well defined. This "production-line" view of the development process assumes that time is

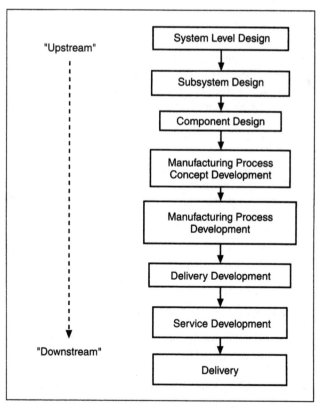

Diagram 2-4. Traditional Product Development

wasted in downstream steps if upstream steps have not yet been completed, with plans solidified.

Although it is true that downstream work must take upstream decisions into account, the major problem not dealt with in the traditional model is that upstream steps may arrive at results that are unrealistic, impractical, or not optimal for downstream implementation.

For example, the engineer may unwittingly choose a design that is unnecessarily difficult to manufacture, or that is expensive to repair in the field. Manufacturing and field support interaction with the designers could influence those decisions and thereby lower the company's downstream costs.

The best approach to making upstream *and* downstream decisions is an interactive one, in which representatives from all functions collaborate, sharing their decision-making processes throughout. In such a dynamic, give-and-take scenario, realistic decisions can be made that achieve the best, most workable results for all functions.

Process and software development processes strongly resemble, or can be made to resemble, product development processes, except that the manufacturing steps refer to the development of tangible objects that support, but are somewhat incidental to, the intangible software or service. These tangible objects might be computer diskettes or computer user manuals in the case of software. For services, such as a credit card service or a home appliance repair service, the tangible objects might be signs, leaflets, or tools that help the service organization to deliver its service.

In some cases, development of these tangible objects may be straightforward, risk-free, or somehow separable from the main development work. In other cases, careful examination of customer needs may reveal that the tangible objects are intrinsically bound to the success of the service.

The Concurrent Engineering Model aims at starting all development process steps as early as possible, even simultaneously (Diagram 2-5). Its success comes from *each step influencing the other* as the development process moves forward. With sufficient communication between people responsible for each step, practical and optimal results are more likely for all steps.

The method by which communication can occur frequently enough and at a detailed enough level to achieve these desirable results, is to treat the people responsible for each development step as a single "multifunctional" team, located together, and all with the same objective—success of their jointly developed product. More and more U.S. companies are adopting this style of product development. It is often referred to as "project," rather than "functional," organization. The objectives of such "project" organization are

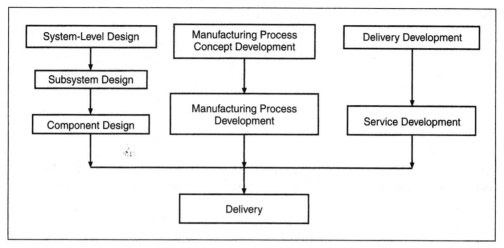

Diagram 2-5. Concurrent Processes

- Earlier start of each development step, leading to earlier completion of all steps
- Optimization of decisions at every step through dialogue and collaboration by representatives from all disciplines, leading to optimal product design and low production and delivery costs
- Overall lower development cost

All these product-development advantages are equally desirable for development of software or services. The competitive pressures are no different. The need to produce software or services rapidly, at low cost, and optimized for customer satisfaction is just as urgent as for tangible products. Hence, the principals of Concurrent Engineering are equally applicable, even if somewhat less familiar to professionals in the software and service domains.

QFD is a central tool in support of Concurrent Engineering. It brings the multi-functional team together in the first place, to develop the top-level House of Quality matrix. At each step in the process, it helps keep the team focused on customer satisfaction, the primary ingredient of product success.

QFD further provides the team a method for prioritizing development actions, which in turn leads them to concentrate most on the actions that will have greatest impact on customer satisfaction. This optimization strategy is a key to reduced development costs. Without an effective optimization strategy, the team would be at risk of optimizing product characteristics that are unimportant to the customer, and thereby raising costs.

In product and manufacturing process design, a key optimization tool is Taguchi's Robust Design method. QFD's prioritization capabilities help development teams decide where to apply Taguchi's methods. In service and software development, Taguchi's methods have not been widely used, although ingenious practitioners have had some successes. Other optimization techniques, such as the basic tools for Statistical Process Control and the Seven Management and Planning Tools, are available to developers of all products and services. QFD provides a road map for application of these powerful tools.

2.8   Kano's Model

The Japanese TQM consultant Noriaki Kano,[5] has provided us with a very useful model of customer satisfaction as it relates to product characteristics. We'll be using the term *characteristics* in a fairly precise way later on, and we'll be drawing sharp distinctions between *customer needs* and *product characteristics*. For now, however, we'll use the term *characteristics* to refer to features or capabilities of a product.

Kano's model divides product characteristics into three distinct categories, each of which affects customers in a different way. The three categories are

  · Dissatisfiers, also known as "must-be," "basic," or "expected" characteristics
  · Satisfiers, also known as "one-dimensional" or "straight-line" characteristics
  · Delighters, also known as "attractive" or "exciting" characteristics

The horizontal axis in Diagram 2-6 shows the actual performance or state of physical fulfillment in delivering each of these product characteristic categories to the customer, while the vertical axis indicates the customer's level of satisfaction.

A competitive strategy for developing products and services must take into account these three categories of product characteristics. It must determine what the current levels of satisfaction are for each of these categories, and it must decide what proportion of project resources to allocate to product or service characteristics in each of the categories.

2.8.1   Dissatisfiers

A dissatisfier is a product characteristic that the customer takes for granted when it is present, but that causes dissatisfaction when it is missing. Dissatisfiers are things that customers don't normally ask about, because they *expect* them to be taken care of. Dissatisfiers are the absence of "expected quality," in the sense that customers expect products to be essentially flawless, and if they are not, the customers are dissatisfied. Examples of dissatisfiers are scratches or blemishes on the surface of a product, broken parts, missing instruction

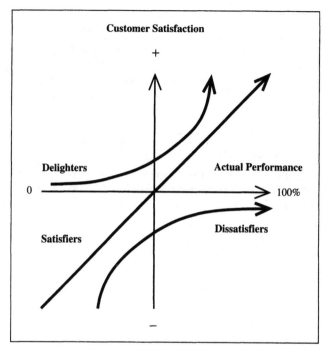

Diagram 2-6. Kano's Diagram

booklets, or missing features that are routinely supplied in similar products. Customers don't tell us they want "expected quality" because they take for granted that we will provide it.

If we deliver a product or service that has many dissatisfiers, customers will be extremely unhappy. Obviously we will have happier customers if we can eliminate these dissatisfiers. However, if we eliminate them all, we won't have achieved a very high level of customer satisfaction. Our customers will hardly notice all the work we've done to eliminate the dissatisfiers—they will just start noticing other aspects of the product or service.

In other words, our best possible actual performance in reducing dissatisfiers can only raise customer satisfaction to a "not dissatisfied" state, but no further.

Although customers won't ask for expected quality, they will be dissatisfied if they don't get it, and they will tell us by complaining. Thus, *customer complaints* are a primary source of information on existing dissatisfiers in our current products.

Many traditional product quality programs center around complaint management. In fact, quality control is sometimes erroneously defined as a process to

Expected Quality	Dissatisfiers
Smooth surface	Scratches, blemishes
All parts work	Broken parts
Product comes with instructions	Missing instruction book
Products of this type normally perform function X	Function X not provided
Product is safe to use	Product is unsafe
Product conforms to local standards	Product is nonconformant

Diagram 2-7. Dissatisfiers and Related Customer Needs

respond to, and perhaps eliminate, customer complaints. The Kano model shows us that this approach is not enough.

First of all, it only deals with expected quality, ignoring the other two categories of quality, Satisfiers and Delighters. Second, merely eliminating Dissatisfiers cannot result in competitively high levels of customer satisfaction.

2.8.2 Satisfiers

A satisfier is something that customers want in their products, and usually ask for. The more we provide of a satisfier, the happier customers will be. Satisfiers are sometimes called "desired quality" because they represent the aspects of the product that define it for the customer. Examples of satisfiers are increased capacity, lower cost, higher reliability, greater speed, and easier use (Diagram 2-8). Satisfiers are the attributes that tend to be easy to measure, and therefore they become the benchmarks used for competitive analysis. In other words, you can expect the satisfiers to be present in all the competitive products, to a greater or lesser extent.

2.8.3 Delighters

Delighters are product attributes or features that are pleasant surprises to customers when they first encounter them. However, if Delighters are not present, customers will not be dissatisfied, since they will be unaware of what they are missing. Delighters are sometimes called "exciting quality" or "unexpected quality." A typical customer reaction to a delighter is to say a to friend, "Hey! Take a look at this!"

As with dissatisfiers, customers don't tell us they want delighters, but for very different reasons. Customers expect "expected quality," so they don't bother to mention it, but customers can't expect "exciting quality" or "unexpected quality" *by definition*. Hence, we cannot learn about product delighters by directly asking our customers.

Desired Quality (Customer Need)	Performance Measurement	Direction Of Goodness
Capacity	Cubic feet of storage	Larger the better
Price	Dollars	Smaller the better
Reliability	Mean time between failure	Larger the better
Speed	Transactions per second	Larger the better

Diagram 2-8. Examples of Desired Quality

Examples of delighters are not as instructive as examples of Satisfiers and Dissatisfiers. Each delighter is unique, and as a group there are no patterns. Some delighters are entire products that have created new markets (a typical consequence of building delighters into products). One very famous delighter is the Sony Walkman. Before its introduction, who could have told you that they wanted a play-only portable cassette player without speakers? After its introduction, everyone wanted one, and a new market was created.

The 3M Post-it Note is another example of a delighter. It's a product that filled user needs that had not previously been filled satisfactorily.

Other delighters are more subtle. They may not have created entire new markets, but only created temporary competitive advantages. Some of these are spare change holders and soft drink holders in automobiles; one-touch recording controls on some VCRs; "redial" buttons on telephone receivers; and graphic user interfaces (GUIs) available on an increasing number of desktop computers.

The needs that delighters fill are often called "latent" or "hidden" needs, either because they cannot be explicitly identified or because customers don't say that the needs are important to them. Some indirect methods for identifying latent or hidden needs and product delighters are discussed in the QFD Handbook part of this book (Chapter 16.) These hidden needs are sometimes intimately linked to customer's perceptions of the limits of technology. As such, they are very difficult to separate from technical solutions.

Let's look at an example. In the year 1840, in the United States, coast-to-coast travel was an arduous and dangerous experience. In hindsight, we might say that the need for a rapid (five-hour) travel method, during which a passenger might read a book or watch a movie, existed even in 1840. However, no one was likely to express the need, because the technology to achieve it was unimaginable. Clearly, if a five-hour transportation method from New York to San Francisco had suddenly become available in 1840, it would have to have been classified as a "delighter," even without the movie!

Since no one in 1840 could imagine the possibility of jet aircraft, we must ask the question, Did the need for a five-hour transcontinental transport really exist then? Technically, one might say that if the need for a three-month transport method (covered wagon) existed, then it would be natural that travelers of the day would have articulated a need for a two-and-a-half-month transport method, or even a one-month transport method. However, a five-hour transport method would have been more than four hundred times faster than the prevailing method, so much faster that it would have had to have been considered fanciful. Did the need exist? Would anyone have articulated such a need?

There is no answer to this question that holds true for all customers and all products. Indeed, there is no clear method for discovering delighters that is guaranteed to work in all cases.

One of the disciplines that QFD helps us to maintain is to separate customer needs from technical solutions. So it is consistent with the intent of QFD to search first for customer needs, and only afterwards for technical responses to those needs, including delighters. Whoever is listening to customers must be knowledgeable and vigilant enough to identify customer needs when customers mention them, and to identify brilliant technical solutions when anyone happens to mention them. Clearly, the needs and solutions must eventually be sorted out, so that the technical solutions can truly be evaluated in terms of how they relate to customer needs.

2.8.4   Market Dynamics of Dissatisfiers, Satisfiers, and Delighters

Delighters often create new markets or new market segments, thereby giving their creators a temporary competitive advantage. Once the novelty of a delighter wears off, and the competition includes the delighter or some equivalent solution into their own products, customers begin to *expect* the delighter, since it's available in all competing products. When this happens, the delighter is no longer "unexpected quality"—instead it becomes "expected quality." In other words, delighters become demoted to satisfiers. After awhile, many satisfiers become "expected quality," and customers assume these satisfiers will be included in the product. When this happens, the satisfiers have become demoted to dissatisfiers.

This migration of quality attributes happens all the time with all products. In order to remain competitive, the product or service developer must continually search for new delighters, provide more satisfiers than anyone else, and see to it that no dissatisfiers reach the customer. QFD is an excellent planning tool for sorting out these different qualities and for managing them.

2.9   The Lesson from Kano's Model

There are two major lessons that Kano's model teaches us.

First, all customer satisfaction attributes are not equal. Not only are some more important to the customer than others, but some are important to the customer in *different ways* than others. For example, dissatisfiers matter not at all when they are met, but seriously detract from overall satisfaction when they are not met. In contrast, satisfiers contribute to overall satisfaction linearly.

Second, the old product quality strategy of responding to customer complaints can now be seen to be inadequate. Customer complaints are for the most part linked to dissatisfiers. A quality strategy based solely on removing dissatisfiers can never result in satisfied customers.

Responding to customer complaints can be thought of as a passive, responsive quality strategy. A strategy which will lead to customer satisfaction and to a leadership product or service must be far more proactive. The strategy must be based on a deliberate policy of seeking out customers and potential customers to discover and characterize their needs, both met and unmet. It must aim at breaking old thought patterns and finding creative ways of meeting those needs and exceeding customers' expectations. Finally, it must be based on a clear, reliable way of estimating the efficacy of each potential method for meeting customer needs, so that the best way can be exploited.

Summary

Any organization's most basic goals are to increase revenues, decrease costs, and produce new products rapidly. We have seen how QFD can be used strategically and tactically to help in these endeavors.

With regard to increasing revenues, QFD provides a key tool for linking customer's needs to product or service design.

With regard to decreasing costs, a principal benefit of QFD is to help development teams plan projects well enough to reduce the likelihood of midcourse corrections, the most devastating source of costs in most development projects.

With regard to rapid product development, the most important way to stay competitive is to respond rapidly to changes in the market environment. Rapid response is best achieved by reducing the development time of new products and services. QFD helps reduce development cycle time by reducing implementation errors, by improving communication, and by supporting concurrent engineering.

Also in this chapter, we were introduced to Kano's model of customer satisfaction. Kano's model teaches us that competitive products must not only be free of "dissatisfiers," but must have a generous allotment of "satisfiers" and "delighters." Thus, a competitive product development strategy cannot simply respond to customer complaints, but must actively determine and address customers' needs. Further, while some needs of customers are fairly easy to discover, the needs that are keys to "delighters" are the unmet and sometimes unspoken needs, which require special methods to uncover, and often require engineering brilliance to meet.

Now that we understand why QFD might be important for any organization that produces products or services, we're ready to judge for ourselves how QFD's contributions fit in. Part II provides a detailed description of the most fundamental part of QFD, the House of Quality.

Discussion Questions

What are your organization's key strategies for reducing costs? What role does reducing rework play?

Have you observed midproject shifts in direction? What have been the benefits of such shifts? What have been the disadvantages of such shifts?

How do project members communicate project goals to each other today? What works well, what doesn't? How would communication be different if you were using QFD?

For some existing product or service you have been associated with, which characteristics are "delighters"? Which are "satisfiers"? Which are "dissatisfiers"? Try the same classification for your competition's product or service. How did you decide on the correctness of your classifications? What data do you have about your customers to confirm your classification?

---

[1] W. Edwards Deming, *Quality, Productivity, and Competitive Position*, MIT Center for Advanced Engineering Study, Cambridge, Massachusetts, 1982.

[2] Despite the ability of some movie actors to drive cars while being visually impaired (for example, Al Pacino in *Scent of a Woman*), most readers of this book will be well advised not to try this stunt.

[3] This pun on the word "embarrassment" is dedicated to my good friend and QFD pioneer, Russ Doane.

[4] *Total Quality Development*, by Dr. Don Clausing, ASME Press, 1994.

[5] Noriaki Kano, Nobuhiko Seraku, Fumio Takahashi, and Shinichi Tsuji, "Attractive Quality and Must-Be Quality," *Hinshitsu* 14, No. 2, February 1984. Published by the Japan Society for Quality Control. The author's access to this Japanese-language article was through a translation by Glenn Mazur.

PART II

QFD at Ground Level

CHAPTER 3

Getting Ready for the Details

Introduction

In order to explain QFD, we must first acquaint ourselves with some procedural building blocks of QFD. QFD uses certain problem-solving and planning tools drawn from a set called the "Seven Management and Planning Tools." This chapter introduces those tools and describes how they will be used in QFD. These tools will come up over and over again, just as a hammer or a saw is used repeatedly in building a house (a "physical house"). If you have not seen these tools before, please take time to learn about them here. Even if you *have* seen them before, it might be helpful to read this chapter to see how they will be applied to QFD.

We are all accustomed to the concept of tools. A tool helps us to perform some function more easily than we could without the tool. For example, a hammer allows us to drive a nail into wood more easily than if we had no hammer. We might be able to drive a nail into wood without a hammer—for example, by using some heavy, hard object, such as a stone. But clearly the hammer is the preferred device for this purpose.

Once we have a hammer and have learned how to use it for its basic function, we can apply it to other uses, such as clearing glass slivers from a broken window. In other words, most tools have purposes beyond those for which they were originally designed. In fact, we take for granted the way in which our familiarity with a particular tool allows us to perform a wide variety of tasks. Tools such as the hammer may be thought of as *mechanical tools* in the sense that they make it easier to perform a task, assuming we know what we want to do in the first place.

The world of Total Quality Management has provided us with a variety of tools that assist us in the tasks we perform in the workplace. The TQM tools generally differ from *mechanical tools* in that they are *decision-making* tools. With mechanical tools, we have already decided what we want to do, and we just want to do it quickly and with minimum effort. In contrast, decision-making tools help us decide what to do in the first place.

The word processor is a good example of a mechanical tool. If we know what ideas we want to communicate, the word processor gives a quick, easy way of putting those ideas into a document. But if we haven't decided what ideas we wish to communicate, the word processor will not make writing very much easier for us.

The TQM decision-making tools help us to organize ideas and data, to interpret those ideas and data, and to decide how to act on the interpretation. If mechanical tools are muscle enhancers, then decision-making tools are mind enhancers. This chapter introduces some of these decision-making tools and shows their relevance to QFD.

3.1   The Seven Management and Planning Tools

In the late seventies, a book appeared in Japan, published by the Japanese Union of Scientists and Engineers (JUSE), entitled *The Seven New Tools*.[1] These tools were intended to provide a level of problem solving power in the conceptual domain equivalent to the power of the "Seven Basic Tools" in the process improvement domain.

The Seven New Tools are usually called the Seven Management and Planning Tools in the United States. Lists of the tools and even the actual number of tools vary a bit from one reference source to another, but most of the tools appear in all lists. The following tools are the mainstays of QFD:

- Affinity Diagram
- Tree Diagram
- Matrix Diagram
- Prioritization Matrix

Other tools often included in lists of the Seven Management and Planning Tools are not directly required for QFD, but are helpful for some of the follow-up work. They are:

- Interrelationship Diagram
- Process Decision Program Chart
- Matrix Data Analysis
- Arrow Diagram

All these tools are worth learning. There are good sources available[2] for learning them. Anyone involved in strategic planning of any type will benefit from the use of these tools. For our purposes, we will look closely at the Affinity Diagram, the Tree Diagram, and the Matrix Diagram. The Interrelationship Diagram (sometimes called the Interrelationship Digraph) is described briefly in Chapter 9. We recommend that you learn about the other tools from other sources.

3.2   Affinity Diagram

The Affinity Diagram (Diagram 3-1) is a powerful tool for organizing qualitative information. It provides for a hierarchical structuring of ideas. The hierarchy is built from the bottom up, and the relationships between the ideas are based on the intuition of the team creating the diagram.

The initial ideas in an affinity diagram can come from one of two main types of sources: *internal* or *external*.

Internal ideas are brainstormed by the team developing the diagram. Brainstormed ideas would be appropriate for a team that has no data to begin with. The team may have some initial ideas about the problem they are beginning to solve, but they may not have begun collecting data, and they may not even know what data to collect.

The purpose of brainstorming is to develop a structure of ideas that describes the team's current understanding of some problem area. Having created this structured model of the problem area, the team could then decide which parts of the problem area they want to attack first, and which parts they need to gather more data about.

External ideas are facts the team has acquired. The facts will generally be anecdotes that apply to the problem at hand. This is generally the situation for QFD teams that want to understand their customers' wants and needs.

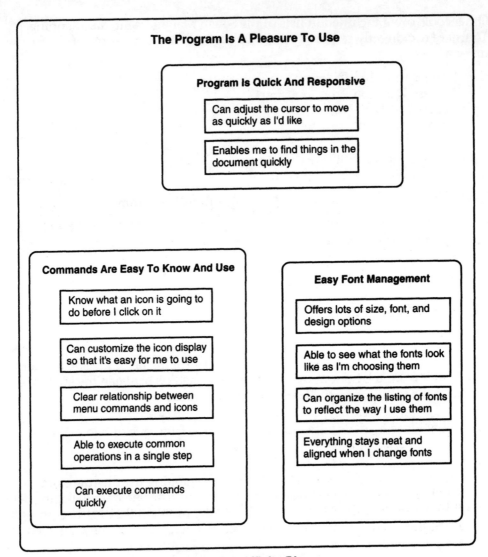

Diagram 3-1. Affinity Diagram

To prepare for QFD, the factual data consists of phrases taken directly from the customer. While customer phrases may not necessarily refer to anything concrete, the customer phrases themselves *are* concrete, since they are direct evidence of customer wants and needs and customers actually uttered the phrases.

Customer phrases, sometimes called *verbatims*, usually come from interviews that developers conduct with customers specifically to prepare for QFD.

However, there are many other possible sources (see a discussion of this in Chapter 16, "Gathering the Voice of the Customer").

To construct an Affinity Diagram, the team writes each phrase on a small card or Post-it Note. The cards are spread on a table, or the Post-it Notes are placed on the wall, where they can all be seen by the entire team. The team then moves the cards around to form clusters that intuitively go together. The use of intuition is an important element of the Affinity Diagram. The use of intuition allows for discovering unexpected relationships, which in turn often spark breakthroughs in understanding the data.

Here is the general procedure for drawing Affinity Diagrams:

- After writing ideas on cards, "scrub" the data. Scrubbing is the process in which each team member explains what he or she wrote on each card, to ensure that the team members all understand it the same way. Often scrubbing helps the team to identify two cards with the same meaning, in which case they are replaced by a single card which best expresses the meaning. Scrubbing also helps reveal cards that have more than one idea on them. These are then replaced by two cards, each of which expresses a single idea.

  For example,

  The system offers lots of fonts, and I can see what each one looks like

  would be replaced by two cards:

  The system offers lots of fonts

  and

  I can see what each font looks like.

- After all cards have been scrubbed, sort the cards into piles that "feel" as if they belong together. There should be no discussion during the sorting process.

  If a card continues to shuttle between two piles because one person wants it with one set of cards, and another person wants it elsewhere, make a duplicate of it and put it in both piles. This rule is helpful during the early stages of sorting. It eliminates the possibility of a test of wills between two team members. In the long run, however, it may be confusing to have a card show up in two places in the hierarchy. Once the team begins assigning titles to the piles (next step), the duplicated cards usually turn out to belong in only one pile.

  Teams often encounter difficulty in "leveling" during the affinity diagram process. This is the problem of ensuring that the cards sorted into a pile are all at the same level of abstraction as each other. For example, consider the following list of ideas:

Can figure out how to turn on the windshield wipers
Can set the car's clock without reading the driver's manual
Can find and use the seat adjustments
The car's adjustment controls are intuitive

All of these ideas could have come from interviews with automobile customers. The first three ideas deal with ease of understanding specific controls in the automobile. The fourth idea,

The car's adjustment controls are intuitive,

deals with *all* of the controls. Since it's more general, it would make a
good title for the other three ideas, and probably for other ideas that
relate to individual controls.

Often the starting ideas in an affinity diagram are at many levels. The
card-sorting and the card-pile-naming processes are the places for noticing these level mismatches and for making the appropriate adjustments.
Some cards move up in the hierarchy, some move down. There are no
foolproof guidelines for determining the right level for each card—level
determination appears to be an art today, not a science. But here are a
few hints that might be useful:

> Keep the number of cards in one pile low. In this way, it's easier to
> compare each card with all the others in order to judge whether or
> not they are at the same level.
> Be on the watch for "world hunger" cards: phrases that are at such a
> high level of abstraction that they are not useful. Typical examples
> of such phrases might be

> **A good product**
> **A world-class solution**
> **Best in its class**

> Look for overly detailed phrases that are at a higher level of detail or
> specificity than the others. These cards probably belong at a lower
> level in the hierarchy.

• After the silent sorting process is complete, discussion begins again. The
team now creates a title card for each pile. The name of the title card
should express the common element in all the cards in the pile. The title
card summarizes the data on the cards in the pile at a higher level of
abstraction. Usually the team discovers in this process that they must
make various adjustments to the composition of cards in the piles. For
example, the following customer needs (for word-processing software)
might have been sorted together during the silent process:

Offers lots of size, font, and design options
Able to see what the fonts look like as I'm choosing them
Can organize the listing of fonts to reflect the way I use them
Everything stays neat and aligned when I change fonts

The team's choice for the title card might have been

Easy Font Management

However, after discussion, the team might decide that

Offers lots of size, font, and design options

is related to

Can Alter The Appearance Of My Material,

but not to

Easy Font Management.

In that case,

Offers lots of size, font, and design options

would be moved out of this pile and placed in another pile, as in Diagram 3-2.

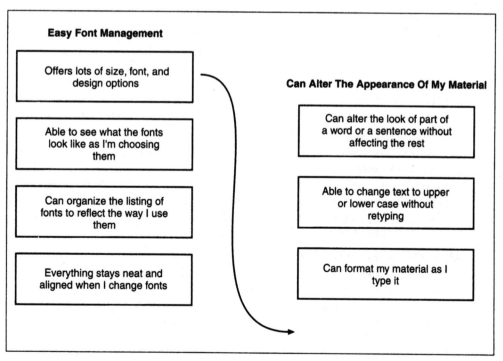

Diagram 3-2. Moving a Card from One Pile to Another

The most common reason for removing cards from piles is that there may be too many cards in the pile. Some teams adopt the rule that there can be no more than ten cards in a pile. If there are more than ten, they assume there must be more than one common theme, so the pile should be broken down into smaller piles. Some teams choose other maximum card counts, such as seven or even three.

· Having named the piles, the teams now sort the *piles* into clusters of piles, according to how *they* intuitively go together. One practical method for doing this is first to place all the subordinate cards in each pile underneath the title card, so they are hidden, and only the title cards can be seen.

Then the team sorts the title cards (and all the hidden cards underneath) as if they were at the level of the first set of cards. The team then creates title cards for these new, higher level piles, just as they created titles for the first level of cards. Each round of sorting and naming results in a set of titles that are summaries or abstractions of the sorted cards. The number of title or summary cards is about one-third to one-tenth as many as the sorted cards. This process of sorting, summarizing, and naming continues until the pile count is lower than the maximum card count per pile.

The affinity diagram method was popularized by quality training organizations such as GOAL/QPC in the early eighties. Later on, another similar method, called the KJ method, was introduced into the U.S.

The KJ method for sorting piles of cards, developed by Jiro Kawakita,[3] probably antedated the affinity diagram and may have provided the evolutionary basis for it. The KJ method stipulates very specific steps for affinity diagramming, including a color coding scheme: cards at the lowest level are written in black ink. Their title cards are written in red. When the red title cards are sorted and grouped, *their* title cards are written in another color, blue for instance.

There are other specific differences between the KJ method and the affinity diagram method. KJ proponents believe strongly in the value of these differences. The KJ method is also strongly linked to a style of problem solving for continuous improvement known as the "WV" method.[4] TQM problem solving in general is beyond the scope of this book, but I urge the reader to gain familiarity with the concept.

· Typical dimensions for affinity diagrams are:

  50 to 150 cards to begin with (tertiary level)
  15 to 25 piles at the first summary level (secondary level)
  5 to 10 piles at the second summary level (primary level)
  3 levels in total

The highest level of abstraction is called the *primary* level, and the other levels are called *secondary, tertiary,* and *quaternary.* It is rare to see an affinity diagram deeper than four levels. This is just as well, because there is no English word meaning "fifth level" to follow the sequence *primary, secondary, tertiary, quaternary*! Most commonly, affinity diagrams are two or three levels deep. The more cards or ideas you begin with, the more levels you are likely to have. Most teams will have difficulties dealing with more than 150 or 200. The author has worked with as many as 1000 cards and does not recommend it.

Diagram 3-3 is an example of an Affinity Diagram of customer needs for a word processing program. The affinity diagram format shown in Diagram 3-1 would be too awkward for the display of so many customer needs, so I have reformatted the needs to the equivalent indented text format. Often teams create affinity diagrams by arranging Post-it Notes on a large wall chart. When the diagram is completed, they transcribe it to an indented text format so it can be printed and distributed to others in a more compact form.

The text indented the least corresponds to the primary level; the next most indented titles represent the secondary level; the most indented text is at the tertiary level.

The customer needs in this example are incomplete, and yet there are a lot of them. The diagram is incomplete in another respect also: The number of customer phrases in some piles exceeds the recommended limit of ten. The hierarchy has been further refined in Diagram 5-4.

Notice how the hierarchical structure, even in its somewhat unrefined state, allows us to "zoom in" or "zoom out" on the data. The primary level headings provide an overview. We can get a more detailed understanding of a primary by examining its secondaries. Each secondary is more fully defined by its tertiaries, and so on.

You may not agree with the organization of the material. There is no "right" organization, nor are there "right" titles, because the relationships and titles are based on subjective judgment. In QFD, we attempt to organize the customer wants and needs the way customers would do it. Although many teams organize the customer wants and needs themselves, it is possible to enlist customers in the process. The additional insights afforded by using customers to organize the data is substantial. See Section 16.5 for a more detailed discussion.

In other applications of affinity diagrams, where the phrases to be organized may come from the team's brainstorming, the "right" organization is whatever makes most sense to the team.

3.3   Tree Diagram

The Tree Diagram, like the Affinity Diagram, is a hierarchical structure of ideas. In contrast to the Affinity Diagram, which is built from the bottom up, based on an intuitive feeling for how the ideas go together, the Tree Diagram is built from the top down, and uses logic and analytical thought processes.

The Tree Diagram usually starts with some already-existing structure—for example, the hierarchy created by the Affinity Diagram process. The team then examines each level of the Tree Diagram, starting with the most abstract or highest level (the primary level), and analyzes that level for completeness and correctness. For example, the list of primaries in Diagram 3-3 is

The Program Is a Pleasure to Use
No Surprises

The Program Is a Pleasure to Use

  Commands Are Easy to Know and Use
    Know what an icon is going to do before I click on it
    Can execute commands quickly
    Can customize the icon display so that it's easy for me to use
    Clear relationship between menu commands and icons
    Able to execute common operations in a single step
    Don't have to read the manual to figure out how to use the program
    Program informs me about all its capabilities and features
    The manual is easy to understand and use
    "Help" function tells me how to do things, not just what things are
    No complicated key strokes to memorize in order to do simple operations
    Don't have to go into Help to understand how to do what I want to do

  Program Is Quick and Responsive
    Can adjust the cursor to move as quickly as I'd like
    Enables me to find things in the document quickly

  Easy Font Management
    Offers lots of size, font, and design options
    Able to see what the fonts look like as I'm choosing them
    Can organize the listing of fonts to reflect the way I use them
    Everything stays neat and aligned when I change fonts

No Surprises

  What I See Is What I Get
    Know what the document will look like when I print it
    Able to see the whole page at once
    Can see what I type as I type it
    Can see all the pages in my document together, side by side
    Able to see subtle spacings easily

Can Control the Shape of My Document

  Can Work With Many Page Styles
    Can create my own document templates
    Can organize my text into tables and charts
    Easy to set up, change, or eliminate headers and footers
    Easy handling of material in multiple columns and rows
    Can use different paper sizes and orientations
    Easy envelope addressing
    Offers me a variety of document types, e.g., letter, invoice, brochure
    Simple to save settings as a default

Can Work With Text and Graphics
Able to mix text and graphics
Able to create charts and pictures in my document
Allows me to easily create presentation style material
Can add pictures/symbols to my document

Can Create and Manage Document Structure
Easy to create footnotes
Able to organize my document as an outline
Easy to create a table of contents
Easy to create an index for my document

Can Modify My Document Any Way I Want To

Can Change the Document Around Any Way I Want To
Easy to move things where I want them in the document

Can Alter the Appearance Of My Material
Can alter the look of part of a word or a sentence without affecting the rest
Able to change text to upper or lower case without retyping
Can format my material as I type it

Can Create Error-Free Documents
Able to easily undo any changes I make
Able to make global reformatting changes effortlessly
The program can proofread my text
Warns me if I'm about to do something wrong—like delete a file
Can check the spelling of words
Makes it easy to edit and correct my work
Won't lose my original text when I type into a section I have highlighted
Warns me if the software is going to bomb
Easy to save my work

Can Get My Ideas On Paper Easily

Can Mix Material From Many Documents
Able to view more than one document at a time
Can combine parts of different documents to form a new one
Enables me to know what is in a document before I open it
Able to move text from one application or document to another
When I bring in documents created in another program, they retain their original appearance
Able to convert document into other types of files (e.g., ASCII)
Can find/view information produced by other programs without exiting
Easy to retrieve and reuse work I created previously

Enhances My Creativity
Helps me quickly capture and save my ideas (e.g., when brainstorming)
Enables me to organize and reorganize my lists

Diagram 3-3. Word Processor Wants and Needs (Partial)

Can Control the Shape of My Document
Can Modify My Document Any Way I Want To
Can Get My Ideas On Paper Easily

Looking at this list analytically, a development team might conclude that the list is incomplete. For example, there is nothing in the list relating to installation, reliability, or interoperability with other software packages. In Diagram 3-3 many needs were intentionally left out, just to make the example a manageable size for this chapter. In real-world projects, there could be real-world reasons why certain areas of customer need are missing. Perhaps these factors didn't show up because of the method used for collecting the Voice of the Customer—limited time for interviewing, or unavailability of certain types of customers to interview, for example. Since the development team knows that these missing factors *ought* to have shown up, they may elect to add them after the Affinity Diagram process is complete.

Likewise, at lower levels, the team may make other additions and amendments, based on their expert knowledge of the subject matter. Working from the top level downwards, the team fills in and completes the hierarchy.

An important caution: when development teams overlay their knowledge on top of customer data, they are assuming some risk. The reason for listening to the customer in the first place is based on the notion that development teams can easily develop misimpressions of what customers' needs are.

In an ideal setting, the data received directly from the customer would be 100% complete and 100% correct. The Affinity Diagram would represent exactly the way customers view the relationships between the data. In that case, there would be no need to modify the data in any way, and the Tree Diagram process would not be necessary.

In real life, the Voice of the Customer process is often incomplete or flawed in one way or another, and development teams often feel the need to "correct" the data. It must be done carefully and with humility, or the whole intent to "listen to the Voice of the Customer" will be undone.

Although the Tree Diagram practice of modifying the hierarchy as the team sees fit must be used cautiously, if at all, with customer data, it can very effectively be used to enhance data that comes from the development team itself—via brainstorming, for example. The Technical Response section of the HOQ is a good example, and we'll be discussing it in Chapter 7. One way of generating the Technical Response (also called the Substitute Quality Characteristics, or SQCs) is for the development team to brainstorm ideas, create an Affinity Diagram, and then complete the Affinity Diagram by applying the Tree Diagram process to it.

When Tree Diagrams are completed they are usually drawn in the form shown in Diagram 3-4. Because Affinity Diagrams and Tree Diagrams are both hierarchical structures, if they are drawn in the same format there is no way to tell whether one is looking at one or the other type of diagram. The difference between them is the method of producing them, not their format. Affinity Diagrams start with the raw data and end with a hierarchical

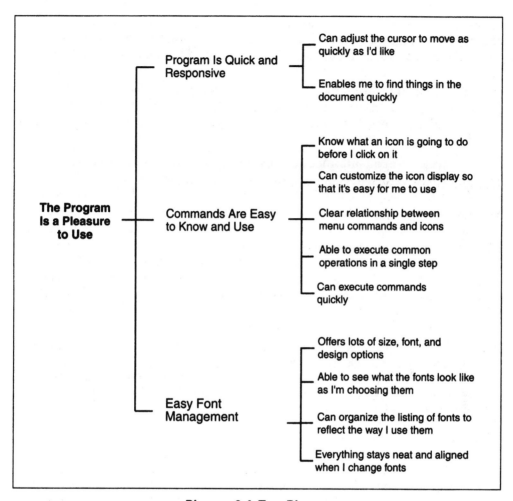

Diagram 3-4. Tree Diagram

structure (bottoms up); Tree Diagrams start with a presumed structure and end up with a detailed elaboration of that structure (top down).

3.4   The Matrix Diagram

The Matrix is a simple but powerful tool that lies at the heart of QFD. Its versatility is heavily exploited throughout QFD. Even though we see matrices in many forms in our everyday work, it's still worthwhile to review the basic concepts here.

A matrix is a rectangular diagram divided into horizontal *rows* and vertical *columns* (see Diagram 3-5). Where a row and a column intersect, we have a *cell*. The cell is uniquely associated with one and only one row–column pair.

We list a range of comparable items along the left side of the matrix. By "comparable items," we mean items that are all attributes or facets of the same general topic. For example, green, red, blue, and yellow are all comparable in that they are all colors. Other comparable items are

- Subsystems of an automobile: suspension subsystem, chassis, steering subsystem, transmission, fuel and ignition subsystem
- Customer needs: easy to learn, easy to use, doesn't endanger me, easy font management

Each of these comparable items is therefore associated with a row of the matrix. We list another range of items along the top and associate each of those with a column. We can use each cell in the matrix to record some relationship between the item associated with the row and the item associated with the column.

In Diagram 3-6, the black circle represents a relationship between "C" and "2." The absence of black circles in other cells indicates that the kind of relationship existing between "C" and "2" does not exist anywhere else in this matrix. This type of entry, which indicates that a relationship exists or does not, is called a *binary* entry in the matrix.

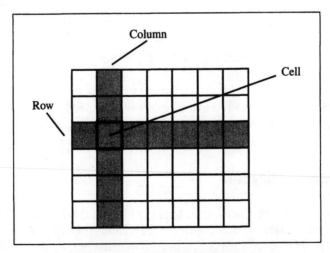

Diagram 3-5. The Matrix Diagram

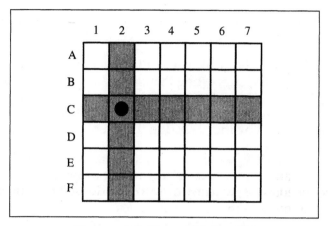

Diagram 3-6. Binary Relationship

In Diagram 3-7 we have a matrix showing several binary relationships between various column and row items. For example, the relationship represented by the black circle exists between items A and 1, between items F and 5, and between items B and 7, as well as between other pairs of column and row items.

Another way of viewing the matrix is to look at entire rows or columns, not just cells. At a glance, we see that the second row shows all the relationships that row item B has to any of column items 1 through 7. There are four circles in this row, so we can quickly see that item B relates to more of the numbered items than any other lettered item except item C, which also has four circles.

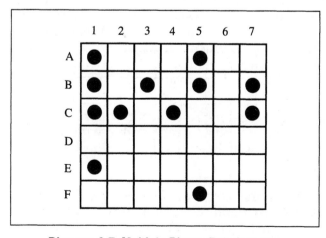

Diagram 3-7. Multiple Binary Relationships

Likewise, column 1 relates to more lettered items than does any other numbered item.

Other more subtle patterns are also evident: most of the relationships are in the top half of the matrix, suggesting that items A, B, and C are somehow different from D, E, and F. Looking from left to right, however, we don't see a very pronounced grouping of any of the numbered items, so in terms of the relationship symbolized by the black dot, there isn't much differentiation between the column items.

The *absence* of relationships is also important in this binary relationship matrix. The column corresponding to item 6 shows no relationship with any of the lettered items, and the absence of circles in the row corresponding to item D also shows no relationship to any of the numbered items. As we'll see later on, this observation is very important in QFD.

3.5   The Prioritization Matrix

The Prioritization Matrix is an extension of the Matrix Diagram. It allows us to judge the relative importance of columns of entries.

We can put many different things into the cells of a matrix. In Diagram 3-7, we used a form of binary data: the black circle or its absence indicated that a relationship existed or did not. Obviously, any other symbol or its absence could have been used. Check marks, X's, dashes, diagonal strokes, and open circles are all common representations of binary relationships.

Besides binary relationships, we can enter numbers, or symbols representing those numbers into the cells. A common QFD practice is to enter numbers that express the strength or degree of the relationship between a column item and a row item. Traditional QFD practice from Japan uses the symbols ◎, ○, △, and blank to indicate "strong relationship," "moderate relationship," "slight or possible relationship," and "no relationship," respectively. Thus, in Diagram 3-8:

- Column item 2 has a strong relationship with row item A
- Column item 7 has a moderate relationship with row item C
- Column item 2 has a slight or possible relationship with row item F
- Column item 5 has no relationship with row item D

Generally we assign numerical values to these degrees of relationship, but many people believe the graphical symbols are easier to see (when the matrix is drawn on a wall chart), and they carry more visual impact than do numbers. Some QFD practitioners prefer the graphical symbols, others the numbers. We'll use both in this book.

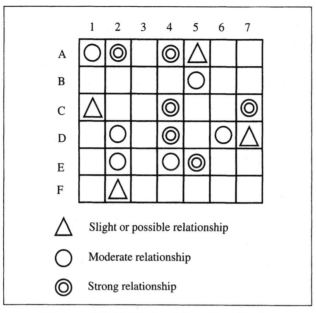

Diagram 3-8. Multivalued Graphical Entries

In Diagram 3-8, we can easily see patterns in the relationships between column items and row items. Row item A has two strong relationships and one moderate relationship with column items; only Row item C also relates strongly with more than one column item. Likewise, we can see at a glance that of all the column items, column item 4 has the strongest relationships with the row items as a group.

The absence of strong relationships can also be observed easily. Column item 3 stands out because it has no entries. Not only has it the weakest set of relationships with the row items, it may be considered irrelevant. In the QFD House of Quality, where the row items are customer needs and column items are technical responses to those needs, an empty column item may indicate an unnecessary technical response, sometimes referred to as a "pet project."

In QFD we assign numerical values to the graphical symbols representing these relationships. The most common numerical values are shown in Diagram 3-9.

If we redraw Diagram 3-8, substituting these values for the graphical symbols, we get the matrix in Diagram 3-10.

The numbers below the matrix are the sums of the strengths of the relationships in the columns. For example, the number 30 at the bottom of column 4

Graphic symbol	Numerical values representing strengths of relationships
◎	9 (less common: 10, 7, 5, 3)
○	3 (less common: 2)
△	1
(blank)	0

Diagram 3-9. Common Relationship Values

represents the aggregate strength of the relationships that column item 4 has with all the row items. It's the largest of the sums, and in QFD our interpretation is that column item 4 is the most important of the column items, in terms of its relationship with all the row items.

The practice of adding the strengths of relationships, as in Diagram 3-10, gives us a good estimate of the overall relationship of the column items with the row

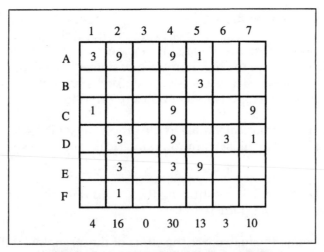

	1	2	3	4	5	6	7
A	3	9		9	1		
B					3		
C	1			9			9
D		3		9		3	1
E		3		3	9		
F		1					
	4	16	0	30	13	3	10

Diagram 3-10. Multivalued Numerical Entries

items, *assuming that the row items are equally important.* Usually the row items are *not* equally important. In QFD we usually associate numerical *weights* or *priorities* with items in a list. When we enter such weighted items in a matrix, we can combine those weights with the relationship values in the cells of the matrix, in order to more realistically estimate the importance of the column items. Let's take a look at this in Diagram 3-11.

In Diagram 3-11, we have added a column of numbers representing the relative importance of each of the row items. (For now, let's not worry about how we arrived at relative importance figures.) We have divided every cell into two parts, separated by a diagonal line. Above the diagonal is the original number (from Diagram 3-10) representing the strength of the relationship between the column item and the row item. In QFD this is usually called the *impact* of the column item on the row item. Below the diagonal is the product of the strength of the impact and the relative importance of the row item. In QFD terms this is usually called the *relationship* of the column item to the row item.

For example, the strength of the relationship (the impact) between column item 4 and row item E is **3**, shown above the diagonal in the cell corresponding to column 4 and row E. The relative importance of item E is **4** (shown just to the right of item E). Therefore, the number below the diagonal in the cell is **3** times **4**, or **12**. This number combines the strength of the relationship and the importance of the row item into a single number (the relationship). By

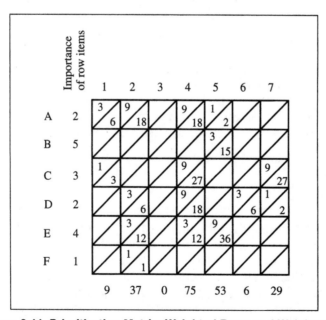

Diagram 3-11. Prioritization Matrix: Weighted Rows and Weighted Cells

multiplying importance times strength, we are expressing the "importance of the strength." To estimate the importance of the column items, we add the "importances of the strengths" of the column items.

We can now look at Diagram 3-10 in a new light. In that matrix, the numbers at the bottom of the matrix are simply the sums of the strengths of the relationships. We did not take the relative importance of the row items into account. Another way of viewing this is that we treated the row items as if they were all equally important. In Diagram 3-11, if all the row items had been equally important, and the importance value had been set to 1, the totals at the bottom of the columns would be the same as in Diagram 3-10. In this sense, Diagram 3-10 is a special case of Diagram 3-11. Diagram 3-11 is called a "Prioritization Matrix." The matrix is translating the priorities of the rows into priorities of the columns. It's the most common type of matrix in QFD.

In QFD, we transfer column prioritization information from one matrix to the rows of another matrix, as in Diagram 1-2. Diagram 3-12 provides a detailed look at how it happens. In the upper prioritization matrix, the row items A through F are the "Whats." In the HOQ, the "Whats" would be the customer needs. The importance of the "Whats" in the HOQ is determined by the development team's work done in the Planning Matrix (as described in Chapter 6). The result of this work is expressed in two Planning Matrix columns: the Raw Weights column (Section 6.6), and the Normalized Raw Weights column (Section 6.7). The information in these two columns is equivalent—the Normalized Raw Weights are proportional to the Raw Weights, but expressed as percentages rather than large whole numbers. Either of these expressions of the Raw Weights are transferred to the columns of the second prioritization matrix (the lower one in Diagram 3-12). They are labeled "Importance of row items" in the lower matrix. The development team generates a new set of "Hows" (denoted by the symbols p through v). The team then determines the impacts of p through v on the column items 1 through 7. These impacts along with the Importances of row items are used to prioritize the column items p through v.

Summary

One of the foundations of Total Quality Management is the widespread deployment of problem-solving tools—mind enhancers—throughout the organization. These tools are usually described as two "tool kits," the Seven Basic Tools and the Seven Management and Planning Tools. While the use of any of the tools is limited only by the imagination and creativity of the user, the Seven Basic Tools, which are mostly numerical tools, are primarily aimed at problem solving and continuous improvement, while the Seven Manage-

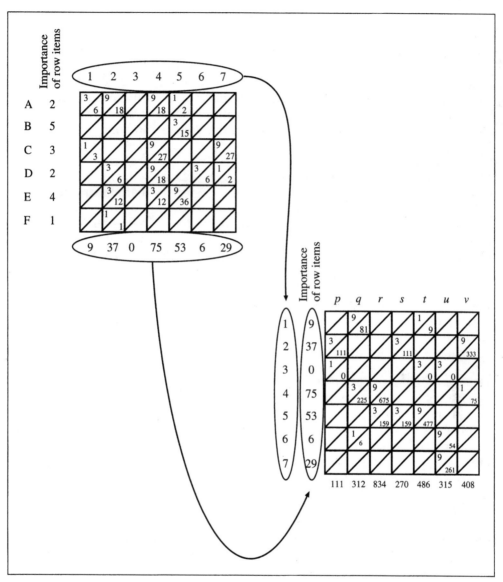

65

Diagram 3-12. Multiple Prioritization Matrices

ment and Planning Tools are intended, as their title indicates, to deal with the types of qualitative information usually associated with planning.

In this chapter, we have looked at the tools most applicable for QFD: the Affinity Diagram, the Tree Diagram, the Matrix Diagram, and the Prioritization Matrix.

The Affinity Diagram is a tool for organizing qualitative data into a hierarchy of ideas. Rather than starting with a preconceived set of categories, the team builds the hierarchy from the bottom up. The first step in QFD involves capturing the Voice of the Customer and using the Affinity Diagram process to build a hierarchy of customer needs, based on the words and phrases of the customer. The hierarchical structure is critical to our ability to understand and manage the complex set of ideas that we call the Voice of the Customer. It allows us to zoom in on specific aspects of the Voice, or to zoom out to understand the Voice at a more abstract level.

The Affinity Diagram is used in other parts of QFD as well, to organize other types of nonnumerical data, such as product or service features and functions.

The Tree Diagram also results in a hierarchy of ideas. Unlike the Affinity Diagram, however, a Tree Diagram's building process starts at the top of the hierarchy, with preestablished categories, and completes and refines the hierarchy by working downward to increasingly detailed levels.

The Affinity Diagram process and the Tree Diagram process are usually used in tandem. First, the Affinity Diagram process is used to create a hierarchy suggested by the raw data; second, the Tree Diagram process is used to complete and refine the structure.

The Matrix Diagram provides a method of mapping relationships between two lists of ideas or concepts. The Prioritization Matrix provides a way of expressing the strengths of those relationships, as well as the relationships themselves, by using numerical values or graphic surrogates of those values in the matrix's cells. Finally, by multiplying weights of the left-hand items by the strengths of the relationships, the development team can calculate weights or priorities of the items along the top of the matrix.

The Prioritization Matrix occurs frequently in QFD. The priorities calculated in one matrix are often transferred to another matrix and become the weighted importance of row items in the new matrix. They are then used, in conjunction with impacts in the new matrix, to compute the priorities of the column items in the new matrix.

Now that we have familiarized ourselves with these tools, it's time to use them by building the House of Quality. Chapters 4 through 11 will take us on a section-by-section, or "room-by-room" tour of the House of Quality.

Discussion Questions

If you have never used the tools described here, take some time to do so before you attempt your first QFD.

Affinity Diagram: What types of concepts or ideas in your work could be organized by means of an Affinity Diagram? Tasks in a project? Elements of a plan? Customer wants and needs? Pick an area of importance to your current work, brainstorm all of its elements, and affinitize them. First do it by yourself, then try it with a small team that has a stake in the subject.

What were the differences between doing it by yourself and doing it in a team?

Were there any surprises? If so, what were they, and how did the Affinity Diagram process contribute to them? If there were none, explain to yourself how using a process you've never used before could give you no new insights.

Tree Diagram: After completing an Affinity Diagram, use the Tree Diagram process to complete the hierarchy. Try it by yourself, then with a team, as with the Affinity Diagram process above. Ask yourself the same questions as with the Affinity Diagram process.

Matrix Diagram: Take the most recent plan that you had any part in creating. List the objectives of the project on the left of a matrix, and list the actions identified in the plan along the top. Fill in the matrix, showing the strength of the relationship between each objective and each action. Based on your examination of the completed matrix, can you think of any way to improve the plan?

---

[1] Shigeru Mizuno, editor, *Management for Quality Improvement: The 7 New QC Tools,* Productivity Press, 1988.

[2] "Memory Jogger Plus+II", by Michael Brassard, GOAL/QPC, 1994.

[3] Shoji Shiba, Alan Graham, and David Walden, "A New American TQM: Four Practical Revolutions in Management," Productivity Press/Center for Quality Management, 1993.

[4] Ibid.

CHAPTER 4

Overview of the House of Quality

Introduction

In this chapter, we'll expand the very brief overview of the House of Quality that was given in Chapter 1. This is still an overview, but it's considerably more detailed than the previous one. In this chapter, we'll introduce each of the sections of the House of Quality, and we'll discuss what information that section contains, and how it is used. Once we've completed this overview, we'll look at each section individually.

The House of Quality is the central construct of QFD. Most of Parts II and IV deal with the House of Quality. Other parts of the book, especially Part V, describes QFD as a process that links the various phases of the development process together, from customer to delivery.

There are two reasons I have chosen to focus so much on the HOQ. First, it contains many of the features we will see in other parts of QFD, so once we have studied it, the remaining QFD matrices and charts will be fairly easy to understand. Second, just about everyone who uses QFD starts with the House of Quality. For better or for worse, many teams choose not to use the rest of QFD. Therefore, a mandatory basic introduction to QFD must include the HOQ, and once mastered, the HOQ leads fairly naturally to various extensions.

If you choose to stop with the House of Quality, you owe it to yourself to know what you have chosen not to do. Please be sure to familiarize yourself with the organizational implications of QFD (in Part III), and with the possible follow-on steps after the HOQ (in Part V). Since we must put first things first, let's turn our attention to the House of Quality.

In most cases, the first matrix a QFD team puts together is the House of Quality. Many teams never create any other matrices (although the use of additional matrices is increasing).

In this chapter we'll present the "textbook" House of Quality. It contains just about everything that anyone has ever put into the HOQ. As you read this chapter, and all of the rest of the book, bear in mind that every aspect of QFD is a candidate for modification or ommission. New QFD elements can and should be added according to the needs of the development team. In almost every application, the HOQ as described here is modified to better help a team solve its particular problem. As we explore more of the details of the HOQ throughout this book, we'll consider the most common variations, and we'll look at the benefits and drawbacks of each.

4.1   Tour of the House of Quality

The House of Quality is a very complex matrix in the sense that it consists of several matrices attached to each other.

Referring to Diagram 4-1, let's a take a quick tour of the House of Quality.

The first section (or "room") of the HOQ to be constructed will almost always be the Customer Needs/Benefits section. The Customer's Wants and Needs are normally derived from the actual words of the customer by any of several methods. We'll be covering many of those methods in detail in Chapter 16. Once gathered, the customer phrases are developed into a hierarchy by means of the Affinity Diagram process, with the most detailed needs at the lowest level, and the more abstract needs in higher levels of the hierarchy, as we saw in Chapter 3.

Most development teams collect the Voice of the Customer from interviews, and then create the hierarchy of wants and needs themselves, although it is possible and desirable to have the customers strongly influence or completely determine the structure. Failure to maximize customer involvement in this process often leads to important misunderstandings of the customers by the development team. When product development teams don't understand their customers' needs well, the product planning activity often gets mired

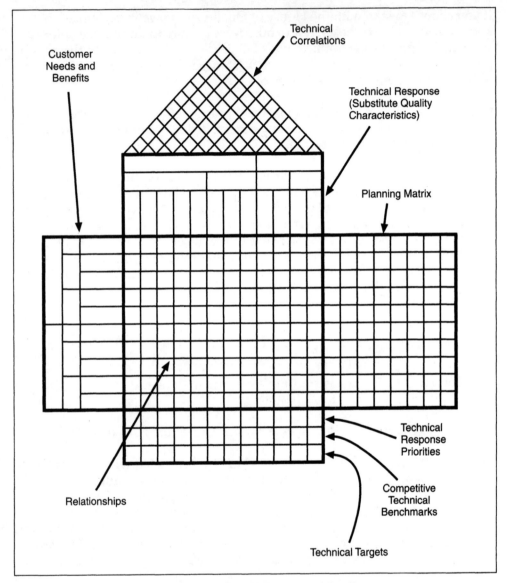

Diagram 4-1. The QFD House of Quality

in confusion. This results in slow product planning and costly midcourse corrections, not to mention noncompetitive products.

The Planning Matrix (sometimes called the Preplanning Matrix) is often the second section to be constructed. Traditionally this section is shown at the right of the House of Quality, as in Diagram 4-1, although as a practical matter it's more convenient to draw it immediately between the Customer Needs/ Benefits and the Relationships. In it, the development team records its answers to a variety of marketing and product planning questions.

The Planning Matrix calls for high-level product goal setting, based on the team's interpretation of the market research data. The goal setting has the effect of combining the company's business priorities with the customer's priorities. It is therefore a crucial step in product planning.

The specific information in the Planning Matrix is (for each customer want/need):

- How important is the need to the customer? (Generally answered by market research.)
- How well does the team's current most similar product or service offering meet customers' needs? (Generally answered by market research.)
- How well does the competition's current most similar product or service offering meet customers' needs? (Generally answered by market research.)
- How well does the team want to meet customers' needs for the product or service being planned? (The team makes this determination.)
- To what extent could meeting a need well be used as a sales point? (The team makes this determination.)

The answers to these questions combine to create a prioritization or rank ordering of the Customer Needs/Benefits.

One reason to fill in the Planning Matrix immediately after the Customer Needs/Benefits are completed is because once the Customer Needs are prioritized, the QFD team may choose to restrict its analysis only to the highest ranking Customer Needs. This considerably reduces the time required to complete the QFD process. If the Planning Matrix were postponed until another time, say after the Relationships section was filled in, the development team would not be able restrict its analysis, since it would not know which Customer Needs were most important to them.

However, some practitioners generate the Technical Responses and even determine the Relationships before developing the Planning Matrix. An advantage to this sequence is that the team will be required to become extremely familiar with the customer needs in order to generate the Technical Responses. Hence, they will be that much better prepared to do the goal setting and high-level analysis in the Planning Matrix when they get to it.

The third section of the House of Quality to complete is the Technical Response section. The Technical Response can be thought of as a set of product or process requirements, stated in the organization's internal language. Sometimes they are called the "Corporate Expectations," to distinguish them from the "Customer's Expectations." There are a variety of different types of information that are placed here. The most common alternatives are

- Top-level solution-independent measurements or metrics
- Product or service requirements
- Product or service features or capabilities

Whichever type of information is chosen, we call it *Substitute Quality Characteristics (SQC)*. Just as the Customer Needs/Benefits represent the Voice of the Customer, the SQCs represent the Voice of the Developer. By placing these two Voices on the left and top of the matrix, we will be able systematically to evaluate the relationships between them.

Whether teams use measurements, requirements, or features as their SQCs depends on the design methodology of their organization. Some organizations prefer a structured design process in which the design is first expressed very abstractly or independently of the ultimate design, and then progressively more concretely in a series of steps sometimes referred to as "stepwise refinement." This approach is particularly favored by software engineers. Other organizations are more comfortable with a concrete design of their product or service at the earliest possible moment. In fact, many organizations conceive of the design before they assess customer or market needs. These differences in style are all compatible with QFD and can be handled by "Static/Dynamic" analysis, as described in Section 12.2.

Regardless of which type of Substitute Quality Characteristic is placed along the top, the amount of detail may need to be managed in much the same way that the amount of customer detail needs to managed. When there is a great deal of detail, the SQCs can be arranged hierarchically by means of the Affinity Diagram process, followed by the Tree Diagram process, as in Diagram 4-2. The hierarchy gives the QFD team some freedom to conduct their analysis at a high or low level of detail, by choosing the Primary, Secondary, or Tertiary level of the hierarchy. The higher the level, the smaller the Relationship section; the lower the level, the more detailed the analysis.

The fourth step is to complete the "Relationship" section of the House of Quality. This is the largest section of the matrix, and therefore represents the largest volume of work. A variety of shortcuts are possible, which we'll discuss later in the book. This step uses the Prioritization Matrix method. For each cell in the Relationships section, the team enters a value that reflects the extent to which the Substitute Quality Characteristic (at the head of the column) contributes to meeting the Customer Need (to the left of the row). This value, along with the prioritization of the Customer Needs/Benefits, establishes the contribution of the Substitute Quality Characteristic to overall customer satisfaction. We'll discuss the details of computing the contribution in Chapter 8. Once the contributions of all SQCs have been computed, the

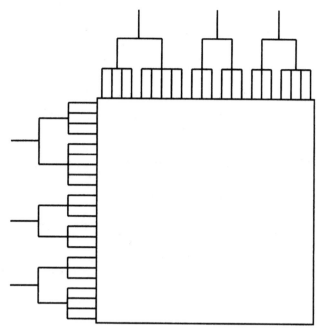

Diagram 4-2. Tree Diagrams on the Left and on the Top

SQCs are essentially prioritized. Those with greatest overall impact on customer satisfaction are most important. This fundamental result is one of the most important outcomes of using QFD.

Some QFD teams, especially those working in less disciplined environments, abandon QFD at this point and use the priorities of the Substitute Quality Characteristics to plan later stages of the development project.

Other teams next use the prioritization of the Substitute Quality Characteristics to provide guidance about further product planning activities. These activities are Competitive Benchmarking and Target Setting, the fifth and sixth steps in completing the House of Quality. They occupy the bottom two lines of the House of Quality, labeled "Competitive Benchmarks" and "Targets" in Diagram 4-1. Competitive Benchmarks and Targets are normally expressed in language compatible with the language of the Substitute Quality Characteristics.

The seventh and usually final step in completing the House of Quality is to fill in the Technical Correlations Matrix (the "roof"). This matrix is used to record the way in which Substitute Quality Characteristics either support or impede each other. This information helps QFD teams to identify design bottlenecks, and it helps them to identify key communication paths among designers.

This completes our overview of the House of Quality; in Chapter 5, we'll begin looking at each of the sections more closely.

Summary

We've just completed a tour of the House of Quality. The HOQ matrix acts as a repository of marketing and product planning information. The key inputs to the matrix are customer wants and needs, product strategy information, and Substitute Quality Characteristics (a high-level formulation of product or service requirements). Other information that can be placed in the HOQ is product benchmarking data and target values.

The HOQ contains many sections or rooms, each of which can and should be customized by the development team to meet its needs. While various sequences for working on the sections each have advantages, the team must consciously choose a sequence and plan its work accordingly.

The most strategic judgments the team must make are in the Planning Matrix. Here the team sets customer satisfaction goals, which have the effect of combining the company's business objectives with the customer's priorities.

Another section of the HOQ (the Technical Response) contains high-level product or service requirements. The Relationships section provides a mapping of how the elements of the Technical Response relate to meeting customer needs, and hence to overall customer satisfaction.

The Correlations section ("the roof") of the HOQ is used to analyze relationships among elements of the Technical Response.

The section below the Relationships contains the priorities of the Technical Response elements, along with a roadmap for competitive benchmarking and target setting.

We're now ready to drop down to another level of detail and discuss the sections of the House of Quality, one by one. The first will be the Customer Needs section.

Discussion Questions

Identify which information normally used for the HOQ was actually used in a recent product planning activity that you participated in. Did you use information beyond the data required for the HOQ? How could you customize the HOQ to include this extra information?

CHAPTER 5

Customer Needs/Benefits Section

Introduction

This chapter describes how QFD represents the Voice of the Customer. In order to understand the VOC representation, we'll discuss how the Voice of the Customer is collected and analyzed. The main steps are

- Listen to the customer and capture the customer's unstructured words
- Sort out the different types of comments the customer has made into various categories
- Take one of the categories, the true customer needs, and create a hierarchical structure of those needs that allows them to be worked with at various levels

This is a very important step in QFD for the obvious reason that the VOC is one of the main inputs to the QFD process. If you believe that "garbage in" produces "garbage out," then you know how important it is to do this phase of QFD right. By the time we've finished this chapter, you'll have a good idea what the VOC should look like.

This chapter deals with VOC basics, which include a single customer segment or category and methods for distinguishing between customer needs and solutions. The chapter also deals with the structuring of customer needs using the Affinity Diagram. Chapter 16 covers more advanced issues, including

dealing with differing types of customer needs and various methods for interviewing customers, some of which are aimed at developing breakthrough
product concepts.

5.1   Gather the Voice of the Customer

The Customer Needs section of the House of Quality (see Diagram 1-1,
Section A, and Diagram 4-1) contains a structured list of needs customers
have for the product or service being planned. This section is usually derived
from the "Voice of the Customer"—literally, statements or fragments of statements made by customers or potential customers.

The usual steps in creating the Customer Needs Section are

1. Gather the Voice of the Customer:
   • Interview customers
   • Gather customer complaints
2. Sort the Voice of the Customer into major categories, including
   • Needs/benefits
   • Substitute quality characteristics
   • Reliability requirements
   • Other
3. Structure the Needs in an affinity diagram
4. Arrange the Needs in the Customer Needs Section

Before gathering the Voice of the Customer (VOC), the team must decide
who the customer is. If, as is often the case, there is more than one category
(or market segment) of customer, the team must decide on the relative importance of the various customer categories and treat them appropriately. All of
this is covered in Section 15.3. At this point, we'll simplify our discussion by
dealing with the treatment of a single customer category.

5.1.1   Interview Customers

The Voice of the Customer is gathered by a variety of methods, all of them
aimed at asking the customer to talk about his or her needs for a product or
service of the type being planned. Some developers implement this step by
conducting a survey in which, for the most part, customers are asked their
opinions on a series of predetermined topics. This is a big mistake, because
the survey designers have no basis for determining the topics to be asked about.

A much better approach is to identify customer needs by interviews developed around open-ended questions. Interviewing and analysis techniques are

described in the QFD Handbook, Chapter 16. The idea is to let customers speak for themselves as much as is practicable.

The result of the interviews is a set of customer phrases representing the customers' wants and needs. Because customers as a rule don't structure their thoughts about product needs, the customer phrases will initially be a mixture of true needs, most favorite and least favorite product features, complaints, suggestions, and other types of comments. All of these comments have potential value during the development process, but only the customers' true needs are needed for the HOQ. Section 5.2 will discuss the sorting of customer phrases.

5.1.2 Gather Customer Complaints

Another source of customer needs, in addition to interviews, is customer complaints. Most organizations have special organizations and processes for handling complaints, since they represent a major nightmare to any company— the nightmare of customer dissatisfaction. Too often, companies regard complaint management as their quality control mechanism. As we have learned from the Kano model, this strategy is not enough to make a company competitive. However, removing dissatisfiers from a product is certainly a necessary, if not a sufficient, step to competitiveness. Hence, it can be useful to include customer complaints in the complete Voice of the Customer.

Typically, companies will maintain databases of customer complaints. These databases can be quite large, and their organization will not normally be convenient for merging into a customer needs affinity diagram. Most complaints are classified by the severity of the complaint or by the type of response required to deal with the dissatisfied customer.

The following is a suggested method for extracting useful VOC information from the complaints:

- Randomly sample a manageable number of complaints from the database, say two hundred to four hundred complaints. Sample from *all* complaints. There is a tendency for customer service departments to discount or ignore certain types of complaints—for example, misuses of the product, requests that are beyond the control of the customer service representatives, and requests for new features. However, these types of comments from customers are extremely important to the developers of new products.
- Using the expertise of developers and customer service personnel, translate the complaints into positive phrases or concepts that represent the underlying customer needs expressed by the complaints.

- Winnow the resulting customer needs phrases by removing duplicates. Maintain an indication with each phrase that it was derived from complaints.
- Merge the resulting phrases in with the phrases acquired by interviews.
- Develop the affinity diagram of customer needs.

Consider restructuring the complaints database so that it can provide more readily available customer input in the future.

5.2   Sort the Voice of the Customer into Major Categories

Because customers often ask for solutions without revealing the underlying need, and because customers' words are not constrained by any particular discipline, the phrases must be sorted before the customer needs can be structured. The standard QFD construct for doing this sort is called the Voice of the Customer Table (VOCT).[1] The VOCT traditionally has two parts.

The VOCT Part 1 (Diagram 5-1) is aimed at capturing the context of customer needs, so that the widest possible range of customer needs is identified and can be understood at a glance. The VOCT is not a matrix; it is a columnar list of customer phrases. The columns are used to provide quick visual clues as to the nature of the data.

The I.D. column identifies the source of the customer phrase. For example, it could contain an interview number, page number, line number, or date of interview. Its purpose is to provide a link back to the source of the phrase in case further information about the phrase is required.

I.D.	Customer Demographic	Customer Need	Use									
			What		When		Where		Why		How	
			I/E	Data	I/E	Data	I/E	Data	I/E	Data	I/E	Data

Diagram 5-1. Voice of the Customer Table, Part 1 (©GOAL/QPC, used with permission)

The Customer Demographic column stores information such as age, income bracket, occupation, or location of the person who provided the data.

The Customer Need column contains the want or need as it appears or will appear in the affinity diagram. Generally, the same need will be expressed in slightly different ways by different customers. The development team will settle on one wording of this need that best represents all the variants.

The Use section holds information that describes what customers do that have implications for the design of the product or service. For example, if the product being designed was a flashlight, the designers would be interested in how customers use flashlights, in order to design a competitive product. Do customers use the flashlight to provide light while changing a car tire in the dark? Do customers use the flashlight as a reading light in bed? These uses probably require very different designs. The VOCT provides a way of listing the range of uses so the developers can make the right design decisions.

The Use section is broken down into several subsections in order to provide a structure for understanding the context of use. A phrase such as

I carry the flashlight in my car in case I need to change a flat tire at night

might result in two data entries: "In my car" would go in the "Where" column, and "To change a flat tire at night" would go in the "Why" column. The phrase:

I only use the flashlight for reading

could yield "beside the bed" in the "Where" column and "for reading" in the "Why" column.

The headers (What, When, Where, Why, How) cover the usual categories of general questions that help interviewers and data analyzers uncover as many aspects of a situation as possible.

Beside each Data column is an I/E column (Internal/External) to indicate whether the data was generated by a developer or company employee (I), or came from a customer (E). The idea behind the I/E column is this: if a customer mentions a specific context for use of the product, the developers may be able to generalize from this specific context to a wider range of contexts. For example, "change a flat tire at night" might suggest:

Take rubbish out at night
Fix a leaky roof at night
Patch broken window at night
Complete an urgent outdoor chore at night

These additional internally generated contexts provide a broader picture of what may be required of the product or service, and they make the requirements more vivid to the developers. However, they should be labeled (I) to indicate that the phrases did not come directly from customers.

The VOCT Part 2 sorts the data a different way (Diagram 5-2).

In the VOCT Part 2, the customer phrases are placed in one list or another depending on whether the phrase is a true customer need, a suggested or requested product function, or any of the other categories the development team may be interested in. Here are some example customer phrases that might be sorted into the VOCT Part 2 categories. These phrases all relate to the use of icons in a word processor.

Customer need:

I don't want to click the icon in order to find out what it does (converted into:) **Know what an icon is going to do before I click on it**

Substitute Quality Characteristic:

Percent of correct identifications of icon meaning by users in a panel test (More Is Better)

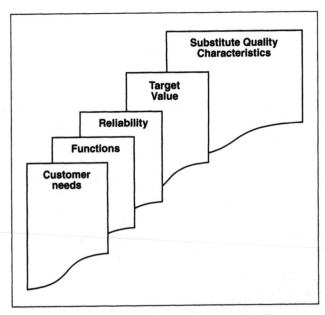

Diagram 5-2. Voice of the Customer Table, Part 2

Function:

> **Icon, when clicked, causes operation to be executed**
> **Icon, when cursor passes over it, causes explanatory message to be displayed**

Reliability:

> **Whenever explanatory message is displayed, it is the correct message**
> **Everything still works well no matter when or how often the user clicks the icon**

Target Value:

> **96% or better correct identifications of icon function by panel test**

Not everyone uses the VOCT—certainly not both parts of it. Nevertheless, the VOCT was developed to deal with some common problems of interpreting the Voice of the Customer, and it is worth understanding. Everyone processing the Voice of the Customer must do at least some of the analysis implied by Part 2 of the VOCT.

Some general guidelines for distinguishing between the main categories in the VOCT Part 2 follow.

5.2.1   Customer Needs

A Customer Need is a statement, in the customer's words, of a benefit that a customer gets, or could get, or might get, from a product or service. Ideally, the benefit should be stated in a way that is independent of the product or service being proposed to meet the need. The word "benefit" is often used in conjunction with "need" to characterize the type of information appropriate for a customer need. "Benefit" shifts the focus toward the customer and away from the Technical Response.

This seemingly simple guideline is much more difficult to apply than you may think. Consider customer needs for an automobile, for instance. Some needs, such as

> **Safe transportation**
> **Comfortable ride**

are reasonably independent of the concept of a vehicle with four wheels, a motor, seats inside a metal body, and the other general characteristics that define an automobile. However, other needs seem to have the automobile as we know it built into the need:

> **Steering wheel that is easy to control**
> **Responsive to my foot pressure on the accelerator pedal**

The interviewer can attempt to uncover more general needs by probing:

> "If it was responsive to foot pressure on the accelerator pedal, how
> would you benefit from that?"

However, sometimes the underlying benefit is too general to be useful in making product decisions.

Unfortunately, we have no hard-and-fast rules that can guide us as to how general or how specific customer needs statements should be. Suffice it to say: general enough to allow for multiple solutions, specific enough to be clearly applicable to the problem at hand.

5.2.2 Substitute Quality Characteristics

Substitute Quality Characteristics (SQCs) represent an abstract description of the product or service in company technical language. We'll be looking at various types of SQCs and various methods for generating them in Chapter 7. Just as customer needs should be completely defined by the customer, SQCs should be aspects of the product or service that are under the control of the development team.

SQCs can be "solution-independent," or they can define the solution by listing its elements. The best product and service developers know that solution-independent descriptions up front provide the best chance for creative solutions later on. Obviously, SQCs that define the solution up front provide no chance for creative solutions.

Most SQCs are generated by developers in response to customer needs. However, customers sometimes suggest SQCs when interviewed for their wants and needs, and we can use the VOCT Part 2 to sort these SQCs into the appropriate column. In fact, SQCs can come from anywhere—the alert developer will always be receptive to suggested SQCs from whatever source.

The VOCT Part 2 column for SQCs could be thought of as a resource into which SQCs can be stored, regardless of their source, and from which SQCs can be drawn later on in the QFD process.

In any case, we cannot control what types of SQCs customers may articulate; we must be watchful for any type. Later on, in Chapter 8, we'll discuss how the team evaluates the importance of the SQCs.

5.2.3 Functions

Functions are descriptions of ways in which the product or service operates. Examples of functions of an automobile are

> **Open window**
> **Speed up**
> **Slow down**

Basic functions of a word processor would be

> **Edit text**
> **Save document to disk**
> **Check spelling of text**

Functions often are made up of subfunctions, which would also be classified in the VOCT Part 2 as functions. Examples are

> **Open window**
> > **Stop window from opening too far**
> > **Hold window in place when open**
> **Edit text**
> > **Move cursor**
> > **Select text**
> > **Deselect text**
> > **Insert characters**
> > **Delete characters**

When we are sorting customers' phrases, we will often encounter mentions of functions and subfunctions. Later on in QFD, we can deal with the varying levels of these descriptions by using the affinity diagram method. This is described further in Chapter 7.

5.2.4 Reliability Requirements

Reliability relates to a sense of confidence that a customer has, or would like to have, in the service or product. It can show up in negative phrases:

> **Warns me if the software is going to bomb**
> **I don't have to take it** (automobile) **for servicing very often**

It can also show up in positive phrases:

> **Keeps working no matter what happens**
> **My data is preserved no matter what happens** (word processor)

In the Kano model, reliability needs are classified as Dissatisfiers. Customers are very dissatisfied when products or services cannot be relied upon, but are not highly satisfied just because they can be relied upon.

These statements give the developers valuable clues as to how customers view reliability, and therefore as to how to plan and test for the reliability of the product or service. Any developer knows that there are an infinity of ways a product or service can fail, and it is impossible to design or test for them all. The Voice of the Customer can help us to avoid those failure modes that matter most to the customer.

5.2.5   Target Values

Target values are indications as to how much of some technical characteristic a customer wants. Any time customers mention numbers in the articulation of their needs, there's a good chance they are providing target values. Examples are:

> **Have as many as ten documents open at once**
> **0 to 60 mph in ten seconds**
> **Won't wait for telephone service more than one minute**

Target values supplied by customers must be considered cautiously. There are many technical issues surrounding target values that customers may not be aware of. The most important issues are methods of measurement, appropriateness of metric, and appropriateness of value.

Methods of Measurement

When a customer requests

> **Have as many as ten documents open at once**

for a word processor, the developer must determine how to count the documents. Users may or may not be aware that even when one of *their* documents is open, the word processing software may have other documents open, including temporary files which may be needed to "undo" user actions (restore the document to its state prior to the last several user actions), a file with user formatting information in it, and a dictionary file.

When a customer requests

> **0 to 60 mph in ten seconds**

for an automobile, the developer must understand what the customer's assumptions are. What type of gasoline is the car using? How many passengers is the car carrying? What is the condition of the tires? If it is a standard transmission automobile, how is the gear shifting to be performed?

When a customer requests

> **Won't wait for telephone service more than one minute**

for a hot-line call, is the customer including dial-up time and connect time in the computation of one minute? Does the minute start when the initial menu of choices starts, or after the customer has made a choice?

These considerations are not likely to be articulated by customers, yet customers may have very definite assumptions about how to count open documents, how to drive a car, and how to measure one minute. At the same time, developers are likely to be accustomed to standard internally defined methods of measuring, which could carry with them very different assumptions.

In short, customer target values should be probed during the interview in order to arrive at the underlying customer need. They should not be taken too literally at any time. We'll discuss techniques for interviewing customers in Chapter 16.

Appropriateness of Metric

Let's consider again the target value statement

Have as many as ten documents open at once

Here the customer has an underlying need, which we can only guess at without further probing. If the developers knew what the customer's underlying need for "ten documents open at once" was, they might be able to meet that need by a solution that did not relate to numbers of open documents at all. For example, suppose the underlying need was

Able to view more than one document at a time

The customer's solution to meeting this need might be

Have as many as ten documents open at once,

because that is the only method the customer knows for meeting the need. However, other solutions might be available to the developer, such as storing key information (the first line of every paragraph, for instance) from many files in a single compressed file that the user can peruse. Perusing a single compressed file may have disadvantages for some users, but the balance of advantages and disadvantages of this solution might be just right for other users. Determination of whether this solution really meets the user need would depend on deeper analysis of the need. Perhaps some type of contextual analysis (see Chapter 16) would be required.

In this case, the metric, "number of documents open at once," probably was not the right measurement: It presumed one solution, precluded other solutions, and obscured the real need.

While the developer must take the customer's proposed metrics seriously, the proposals must be probed to uncover the real needs. Without knowing these needs, the metrics could lead the developer on a fruitless journey.

Appropriateness of Value

Given the likelihood that the customer may have proposed an inappropriate metric, we should not be surprised if the proposed target value for that metric requires careful scrutiny also.

In general, metrics fall into three categories: "The Larger the Better (LTB)," "The Smaller the Better (STB)," and "Nominal the Best (NB)."[2]

"The Larger the Better (LTB)" refers to metrics for which the worst value is zero, and the best value is arbitrarily large. Examples of LTB metrics are mean time between failure, strength of adhesives, shelf life of consumer products, and support strength of architectural columns.

"Smaller the Better (STB)" refers to metrics for which the best value is zero, and the worst value is arbitrarily large. Examples of STB metrics are audio tape noise (hiss) level, heat loss through an insulating layer, or time to resolve a customer problem.

"Nominal the Best (NB)" refers to metrics for which the best value (called the Nominal) is a specific value determined by the situation, and the worst value is arbitrarily greater or arbitrarily smaller than the Nominal. Examples of NB metrics are clothing size, voltage output of a power supply, or temperature of food.

When the customer says

Won't wait for telephone service more than one minute,

the developer must ask which of the metrics categories is being proposed: LTB, STB, or NB. This is helpful in two ways.

First, there is the possibility of exceeding the customer's expectations. The customer's limit of one minute is likely to be what the customer will tolerate, not what the customer really wants. Rather than accept the target of one minute, the developer has the opportunity of exceeding the customer's expectations by generalizing the target value to a direction and providing a product or service that does even better than the customer asked for.

Second, the customer-defined target may be unrealistic. In this case, the developer may be able to produce a competitive product by approaching the customer's target and exceeding the competition's performance on this metric, but without actually meeting the customer's target.

The terms LTB, STB, and NB are like most other QFD terms: they have not been standardized. Some confusion can arise if the development team is not careful. Various alternative terms are displayed in Diagram 5-3.

Notice the duplication of abbreviations LTB and MB in the diagram. It pays to keep the terminology consistent.

5.3   Structure the Needs

After collecting customer needs from many sources, and even after sorting out the Substitute Quality Characteristics and many other items that are not truly needs, there will still be a large, unmanageable list to deal with. In the QFD process, the needs are arranged into an affinity diagram, which is then completed and refined by using the tree diagram process.

Most commonly, the Tree Diagram is three levels deep. If it has more levels, the lower levels are used as definitions and clarifications of the higher levels. For example, in Diagram 5-4, the Tree Diagram is four levels deep in some places ("Intuitive controls" is at the tertiary level, and its subordinates, beginning with "Know what an icon is going to do before I click on it" is at the quaternary level).

If the tertiary level has been chosen for analysis, then the quaternary level, where it exists, can be used to define more fully the associated tertiary terms. The development team chooses one level for their analysis, placing that level against the left edge of the Relationships section of the HOQ (see Diagram 5-5).

Notice that by arranging the customer needs in a hierarchical tree structure we don't lose any detail. However, the hierarchy allows us to manage the information by choosing to work at a particular level.

LTB (Larger the better) MTB (More the better) MB (Maximum best) GTB (Greater the better)	LTB (Less the better) STB (Smaller the better) MB (Minimum best)	TB (Target best) NB (Nominal best)

Diagram 5-3. Variations on MTB, LTB, TB

The Program Is a Pleasure to Use
  Commands Are Easy to Know and Use
    Intuitive Controls
        Know what an icon is going to do before I click on it
        Clear relationship between menu commands and icons
        Don't have to read the manual to figure out how to use the program
        Don't have to go into Help to understand how to do what I want to do
    Controls under my fingertips
        Can execute commands quickly
        No complicated keystrokes to memorize in order to do simple operations
        Able to execute common operations in a single step
    Can customize to suit my working style
        Can customize the icon display so that it's easy for me to use
    Easy to get the information I need
        Program informs me about all its capabilities and features
        The manual is easy to understand and use
        "Help" function tells me how to do things, not just what things are

  Program Is Quick and Responsive
    Can adjust the cursor to move as quickly as I'd like
    Enables me to find things in the document quickly

  Easy Font Management
    Offers lots of size, font, and design options
    Able to see what the fonts look like as I'm choosing them
    Can organize the listing of fonts to reflect the way I use them
    Everything stays neat and aligned when I change fonts

No Surprises
  What I See Is What I Get
    Know what the document will look like when I print it
    Able to see the whole page at once
    Can see what I type as I type it
    Can see all the pages in my document together, side by side
    Able to see subtle spacings easily

Can Control the Shape of My Document
  Can Work With Many Page Styles
    The page styles I need at my fingertips
        Can create my own document templates
        Simple to save settings as a default
        Offers me a variety of document types, e.g., letter, invoice, brochure
    Handles all my document organizational needs
        Can organize my text into tables and charts
        Easy to set up, change, or eliminate headers and footers
        Easy handling of material in multiple columns and rows
        Can use different paper sizes and orientations
        Easy envelope addressing

  Can Work With Text and Graphics
    Able to mix text and graphics
    Able to create charts and pictures in my document
    Allows me to easily create presentation style material
    Can add pictures/symbols to my document

  Can Create and Manage Document Structure
    Easy to create footnotes
    Able to organize my document as an outline
    Easy to create a table of contents
    Easy to create an index for my document

Can Modify My Document Any Way I Want To

Can Change the Document Around Any Way I Want To
Easy to move things where I want them in the document
Able to make global reformatting changes effortlessly

Can Alter the Appearance of My Material
Can alter the look of part of a word or a sentence without affecting the rest
Able to change text to upper or lower case without retyping
Can format my material as I type it

Can Create Error-Free Documents

Protects Me Against My Mistakes
Able to easily undo any changes I make
The program can proofread my text
Warns me if I'm about to do something wrong—like delete a file
Can check the spelling of words
Makes it easy to edit and correct my work
Won't lose my original text when I type into a section I have highlighted

Protects Me Against Its Mistakes
Warns me if the software is going to bomb
Easy to save my work

Can Get My Ideas On Paper Easily

Can Mix Material From Many Documents
Can find material in other documents easily
    Able to view more than one document at a time
    Enables me to know what is in a document before I open it
    Can find/view information produced by other programs without exiting
Can use material in other documents easily
    Can combine parts of different documents to form a new one
    Able to move text from one application or document to another
    When I bring in documents created in another program, they retain their original appearance
    Able to convert document into other types of files (e.g., ASCII)
    Easy to retrieve and reuse work I created previously

Enhances My Creativity
Helps me quickly capture and save my ideas (e.g., when brainstorming)
Enables me to organize and reorganize my lists

Diagram 5-4. Customer Benefits/Needs

Summary

We've covered a lot of material in this chapter.

The Customer Needs/Benefits section of the House of Quality is the starting place for all QFD activities. The most common source of customer phrases representing their wants and needs is the customer interview. Examination of customer complaint data is also useful. Customers' language is not, however, as structured as is desirable for QFD. The developer must be aware that customers will provide data of many types, which the developer must sort, classify, and structure in order to make useful.

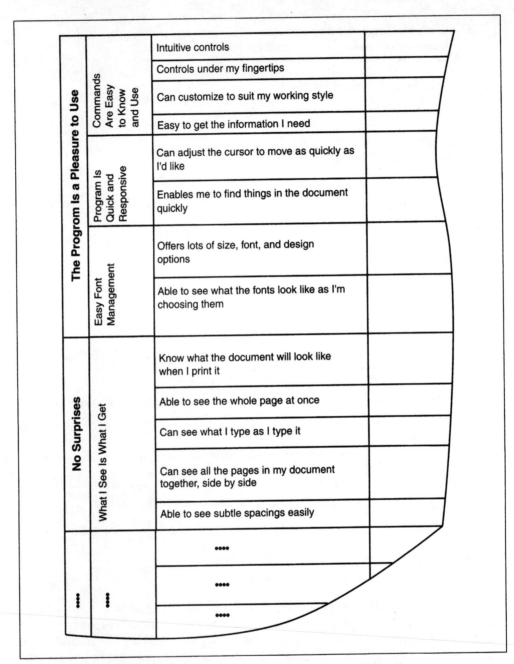

Diagram 5-5. Wants and Needs in the House of Quality

The most common categories used for classifying the Voice of the Customer are Customer Needs, Substitute Quality Characteristics, Functions, Reliability, and Target Values. To make best use of the Voice of the Customer, the developers must record the customer phrases and sort them into categories such as these.

The Customer Needs are the most important category for the early stages of QFD. Because there are usually so many needs, we use the affinity diagram method and the tree diagram method to arrange them in a tree-structured, hierarchical format. The most important parts of this structure are usually the primary, secondary, and tertiary levels of the structure. The structure allows the developers to "zoom in" or focus on the customers' needs at different levels of detail, depending on the appropriate level of analysis. Once the needs are structured, one level, usually the tertiary level, is used for subsequent QFD analysis and is placed on the left of the House of Quality.

Having collected, sorted, and structured the Voice of the Customer, development teams are justified in celebrating! This is an immense, mind-opening task. The number of "ahas" experienced by most teams by the end of this stage is generally dramatic. If "fifty percent of problem solution lies in the statement of the problem," then the development has gotten halfway to a world-class product or service by reaching this point.

The Voice of the Customer we've seen in this chapter is *qualitative data*. In other words, it describes "what" the customer wants. In QFD, we'd also like to have *quantitative* descriptions of the Voice of the Customer. These tell us how important each of the needs is to the customer, and also how well we and the competition are doing in meeting those needs. This quantitative data, along with how to collect it and how to use it to set product strategy, will be described in the next chapter.

Discussion Questions

What data do you have that represents the Voice of your Customer? How was it acquired? How has it been structured? Can you represent it in a tree-structured format? Which customer phrases fit into the categories of "Customer Need," "Function," and "Reliability"? Are there any target values? How do your customers measure these values? What customer needs are the metrics associated with?

---

[1] This diagram appears in many GOAL documents, but in particular "Comprehensive QFD system" by Satoshi "Cha" Nakui, GOAL/QPC, 1991.

[2] An excellent discussion of these terms can be found in Chapter 2 of *Quality Engineering Using Robust Design* by Madhav S. Phadke, Prentice Hall, Englewood Cliffs, New Jersey, 1989.

CHAPTER 6

The Planning Matrix

Introduction

In this chapter we'll show how QFD helps a team to do strategic planning for their product. Just as the Customer Needs/Benefits section is a repository of *qualitative* customer data, the Planning Matrix is the repository for important *quantitative* data about each customer need. The development team will use this data to decide what aspects of the planned product or service will be emphasized during the development project.

This is a long chapter, because the standard, "textbook" Planning Matrix consists of seven very different types of data, each of which must be described separately. The actual number and nature of these seven columns of data is the subject of considerable customization and variation from one QFD to the next. We'll try to indicate what those variations are. I've generally indicated the most common usage for each Planning Matrix column first, with variations afterwards.

The Planning Matrix is the tool that helps the development team to prioritize customer needs. Diagram 1-1, Section B, and Diagram 6-1 are schematic representations of the Planning Matrix. Diagram 6-23 is a worked-out numerical example. The Planning Matrix provides a systematic method for the development team to

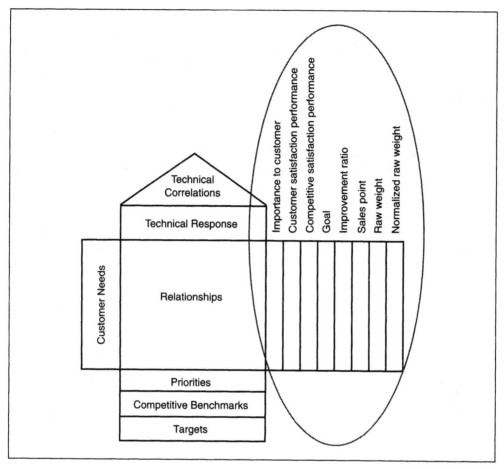

Diagram 6-1. Planning Matrix

- Compare their current product or service performance in meeting customers' needs to the competition's performance
- Develop a strategy for customer satisfaction that optimizes the organization's ability to both sell the product (short-term customer satisfaction) and keep the customer satisfied (long-term satisfaction)

The goal-setting that the development team does in the Planning Matrix will set the tone for the rest of the project.

The Planning Matrix contains a series of columns that represent key strategic product planning information and questions. Sometimes these questions have been called "embarrassing" questions because many organizations either don't know the answers or cannot agree on the answers. Because QFD forces teams systematically to ask all the questions, their level of knowledge about their own product and the competition can become very obvious. The data placed in the planning matrix allows the development team to make strategic decisions about the product or service they're planning. Therefore, the Planning Matrix is extremely important.

The Planning Matrix asks the following key questions for each customer need:
- How important is this need to the customer?
- How well are we doing in meeting this need today?
- How well is the competition doing in meeting this need today?
- How well do we want to do in meeting this need with the product or service being developed?
- If we meet this need well, could we use that fact to help sell the product?

Let's take a look at the Planning Matrix columns and their associated questions one by one.

6.1   Importance to the Customer

The Importance to Customer column (Diagram 6-2) is the place to record how important each need or benefit is to the customer. Three types of data are commonly used in this column: Absolute Weight, Relative Weight, and Ordinal Importance.

6.1.1   Absolute Importance

The Absolute Importance entries are usually chosen from a scaled selection of importances. The number of points on such a scale has been known to range from three to ten. We'll base our examples on a five-point scale where the values 1 to 5 may be defined as

1   Not at all important to the customer
2   Of minor importance to the customer
3   Of moderate importance to the customer
4   Very important to the customer
5   Of highest importance to the customer

Absolute importance values are usually obtained by a survey in which respondents are asked to rate the importance of each need on a scale provided by the interviewer (or described in the survey form). Surveys of this type are often designed and implemented by the development team, without the help of professional market research firms. In some very low-budget QFDs, the

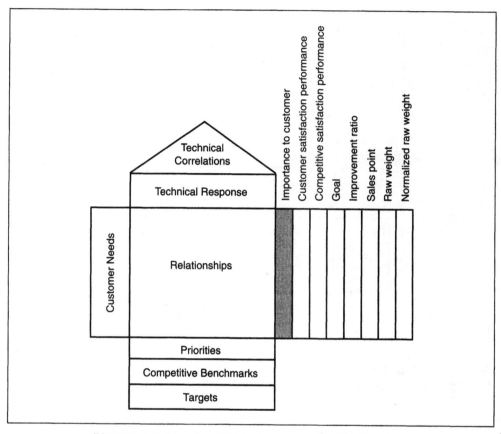

Diagram 6-2. Planning Matrix, Importance to the Customer

product development team will estimate these ratings themselves, or with the help of a few customers—a dangerous and risky undertaking, but not unheard of.

Even assuming that accurate and representative data in an absolute scale is available, there is still a problem using absolute importance: Customers tend to rate almost *everything* as being important (as in Diagram 6-3). While everything *is* important to the customer, the development team is still forced to make trade-offs, because of constrained resources. In addition, customers can select products that do not meet their needs equally well, and still be satisfied overall with them. Thus, if customers can clearly differentiate the importance of different customer needs, the QFD process can help the developer translate those differences into prioritized technical responses. If the absolute weighting data tends to be bunched near the highest possible scores, it does not contribute strongly to helping developers to prioritize.

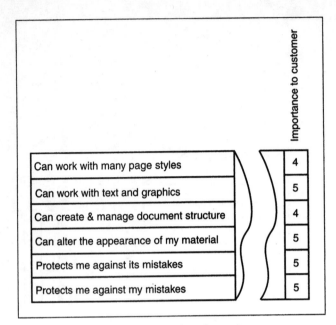

Diagram 6-3. Absolute Importance

A better method for measuring Importance to the Customer uses Relative Importance, which unfortunately requires skills not usually found in development teams.

6.1.2   Relative Importance

The Relative Importance entries reflect that if one need is twice as important as another to the customer, then the importance score of the more important need would be twice the score of the less important need.

Relative importance values are typically placed on a 100-point scale or on a percentage scale. The number 100 indicates the highest possible importance to the customer. Not every customer in a survey will assign the same weight to each customer need, so it is unlikely that any customer need will ever be scored at the maximum value, 100. Typical ranges of relative importance scores are from about 40 to 85.

Diagram 6-4 shows typical Relative Importance scores.

Relative Importance (sometimes called "ratio-scale importance") is measured by asking customers to compare the attributes to each other and indicate importances. There are many methods.[1] One widely used technique presents

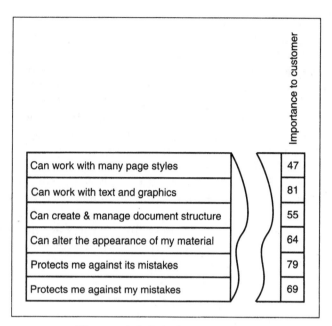

Diagram 6-4. Relative Importance

the customer attributes to respondents in pairs, and asks the respondent to indicate how much more important one member of each pair is compared to the other. This method is called "constant sum paired comparisons." Each attribute is compared to every other attribute in such pairwise comparisons. Information reflecting these choices is stored in a cell in a matrix. The matrix is processed in a manner similar to the Analytical Hierarchy Process method (see Section 15.3.3). The analysis results in weights for each customer attribute that indicate relative importance of the needs.

Any such process of pairwise comparisons carries with it the risk of inconsistent judgments. That is, there is nothing that would stop a respondent from stating that

> A is more important than B
> B is more important than C
> C is more important than A

This type of circular reasoning cannot be easily avoided in surveys, although it can be tested for after the fact. Nevertheless, if the survey process can be constructed in a way that guarantees that inconsistent judgments won't be made, there is no reason why the resulting weights would not reasonably represent the way people feel about the relative importance of their choices. Again, this is work best done by professionals.

Another widely used technique presents the customer with a complete list of possibilities and asks the respondent to arrange these in ascending or descending order of importance. Optionally, the respondent may be asked to assign numerical values representing degree of importance in the sorted list. (The Vocalyst process, described in Section 16.5, uses this technique.) This process has the advantage over pairwise comparisons of assuring consistency. A disadvantage of this process is that certain methods of collecting such data are impractical. For example, if a telephone survey is conducted, the respondent will probably not be able to visualize more than about seven attributes.

Every method for measuring and computing importance is nothing more than a mathematical model of how numbers of people feel. This author finds relative importance the most useful measure of importance for QFD. However, the Relative Importance models are generally complex enough to justify the use of professional market researchers. A variety of statistical and data acquisition decisions must be made during the design of this type of research, and it should not be attempted by amateurs.

6.1.3   Ordinal Importance

Ordinal Importance, like Relative Importance, is an indication of order of importance. Unlike Relative Importance, which indicate *how much* more or less important one attribute is compared to another attribute, Ordinal Importance indicates only *that* one attribute is more or less important than another. Typical Ordinal Importance scores might look like those in Diagram 6-5.

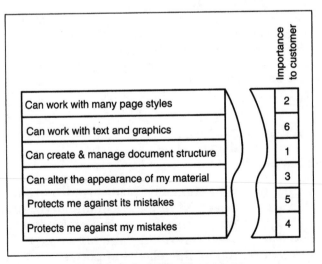

Diagram 6-5. Ordinal Importance

The highest number in the Ordinal Importance column indicates the attribute that is most important to the customer. Most people use Ordinal Importance numbers just the opposite way: the number "1" indicates most important, and higher numbers indicate lower importance. However, QFD arithmetic always treats higher numbers as more important.

Typical methods for measuring Ordinal Importance involve surveying customers and asking them to rank-order the customer attributes, or to assign importance numbers to the attributes as with Absolute Importance. Various forms of averaging of the responses can be used to arrive at Ordinal Importance. For example, if customers are asked to rank-order importance:

- Assign the number 1 to the lowest-ranked customer need on each survey response. Assign ascending numbers to each higher ranked customer need on each response. The highest number assigned (to the customer need judged most important in a survey) would be equal to the total number of customer needs ranked by the customer.
- Add the assigned numbers for each attribute in each survey response.
- The customer need with the highest score will be most important. The customer need with the next highest score will be next most important, and so on.
- The scores *are not* proportional or weighted importances, since the customers were not asked to provide any information about relative importance.

The resulting Ordinal Importances are reasonable estimates of the way the customers surveyed felt. Even though the development team cannot be very confident of *how much* more important one customer need is than another, they can be confident that items at the top of the list are more important overall than items at the bottom of the list.

To compute Raw Weights (Section 6.6), the Importance value will be multiplied by other values in the Planning Matrix. There are two issues to be aware of in this regard:

- It is not strictly valid to multiply an Ordinal Importance value by the proportional values used elsewhere in the Planning Matrix. Nevertheless, this practice is fairly common in QFD, and succeeds in giving the development team a fair indication of what the priorities are in their project.
- The range of ordinal numbers, used as multiplier values, is extremely wide compared to the ranges we see for Absolute and Relative Importance values. For example, with fifteen customer needs, the ratio of largest to smallest ordinal numbers will be 15 to 1. The ratio of largest to smallest Absolute Importance is theoretically 5 to 1, but in practice it is usually about 5 to 3, or 1.6 to 1. The ratio of largest to smallest Relative

Importance is theoretically 100 to 1, but in practice it is usually about 85 to 40, or 2.1 to 1.

Thus, the Ordinal Importance scale, when multiplied by other values in the Planning Matrix, tends to make the highest Raw Weights much larger than the lowest Raw Weights, thus emphasizing the most important customer needs far more than the least important ones.

6.2   Customer Satisfaction Performance

The Customer Satisfaction Performance (Diagram 6-6) is the customer's perception of how well the *current* product or service is meeting the customer's needs. By the current product we mean that product or service currently being offered or delivered that most closely resembles the product or service

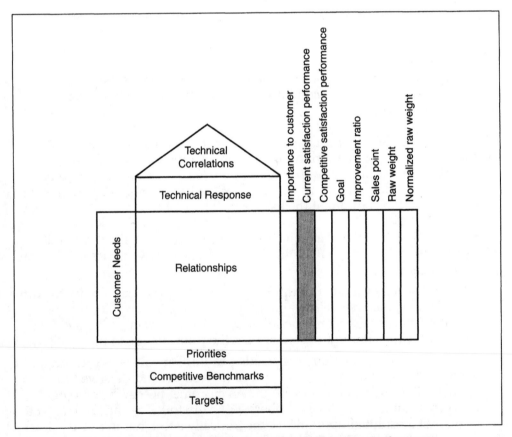

Diagram 6-6. Planning Matrix, Customer Satisfaction Performance

we plan to develop. The information is placed in the column entitled "Customer Satisfaction Performance."

The usual method for estimating this value is by asking the customer, via survey, how well he or she feels the company's product or service has met each Customer Need. This satisfaction level is usually expressed as a "grade" or a performance level. Grades are usually given on a four-, five-, or six-point scale, although sometimes scales up to ten points are used. Often, customers are asked to supply letter grades (A through F) because of their similarity to grades given in U.S. schools. The grades would be converted to numbers, where the highest performance would be translated into the highest number. Typical survey questions for measuring satisfaction levels are shown in Diagram 6-7.

The respondent is expected to answer each question by checking or circling one of the responses. Allowance is often made for respondents who don't know the answer, or for whom the question does not apply.

One method of using data from surveys like this in QFD is to assign numerical values to the possible responses and then compute weighted averages. The weighted average in this case is

	Very poorly	Poorly	Neutral	Well	Very well	Does not apply
How well has the service met your need for courteous telephone operators?						
How well has the word processor met your need for control over appearance of your document on the computer screen?						
How well has the product met your need for minimizing the time it takes to recharge the battery?						

Diagram 6-7. Satisfaction Survey

$$\text{Weighted Average Performance} = \frac{\sum_i [(\text{Number of respondents at performance value } i) \bullet i]}{(\text{Total number of respondents})}$$

Diagram 6-8 illustrates a worked-out example.

The Weighted Average Performance Score for a particular question could be the value we use in the House of Quality Planning Matrix. If almost all customers answered a question with the same value (as in Diagram 6-9), then the reaction of the customer base would be homogeneous with respect to this customer need. However, there is the possibility that the answers may not cluster around a single value. Take a look at Diagrams 6-9, 6-10, and 6-11. For all three cases the Weighted Average Performance is very close to 3, but in Diagram 6-10 and Diagram 6-11 a large percentage of the customers' performance ratings are well above and below 3. In other words, the Weighted Average Performance value is not representative of many of the customers and should be used with caution.

Charts like Diagram 6-10 and Diagram 6-11 may indicate a segmentation of the customer base. By segmentation, we mean that the needs of, or selling opportunities to, a substantial proportion of the customers are different from those of the other customers. They are not being satisfied completely by the existing product. If we want to satisfy most of the customers, the developers may have to develop a technical solution different from the current one.

The main point is that the distribution of customer performance responses to a survey question must be understood before the QFD team blindly repre-

Customer attribute: "Offers lots of size, font, and design options"	Performance / grade	Number of respondents	Performance weight [(Number of respondents) • (Performance)]
Very poorly	1	50	50
Poorly	2	157	314
Neutral	3	626	1878
Well	4	180	720
Very well	5	40	200
Totals		1053	3162
Weighted Average Performance Score (3162 divided by 1053)			3.0028

Diagram 6-8. Weighted Average Performance

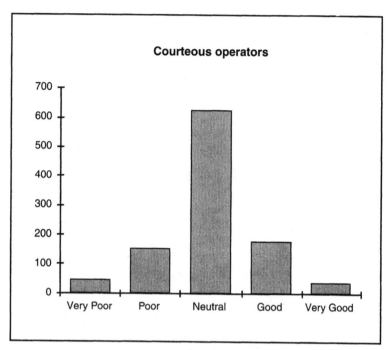

Courteous operators

Diagram 6-9. Homogeneous Customer Performance
(by far, most customers responded with "Neutral")

sents the performance level of all the customers by a single number such as the Weighted Average Performance.

6.3   Competitive Satisfaction Performance

In order to be competitive, the development team must understand the competition. This may sound simplistic, but many development teams do not study their competition very carefully. Because it is usually much harder to reach the competition's customers than their own customers, development teams often operate in the dark with regard to their competition's strengths and weaknesses.

These teams sometimes try to rely on trade journals' evaluation reports for comparisons of their products or services with the competition's. Since the trade journals' criteria for comparison are unlikely to match the customer attributes that a development team has created from its own customer interviews, the trade journal comparisons are very difficult to use in these side-by-side comparisons.

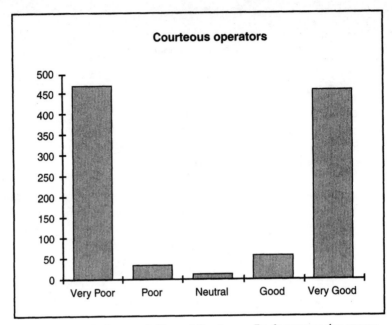

Diagram 6-10. Segmentation of Customer Performance (as many customers responded with "Very Poor" as responded with "Very Good")

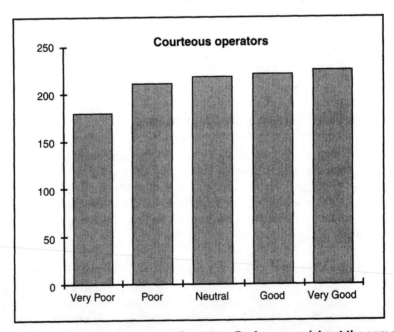

Diagram 6-11. Segmentation of Customer Performance (about the same number of customers responded with each of the five possible responses)

QFD provides a method by which the development team can record the competition's strengths and weaknesses alongside its own. The comparison can be shown at two important levels: first, in terms of Customer Needs, and second, in terms of Technical Response (Substitute Quality Characteristics). In the Planning Matrix, the development team has the opportunity to compare, side-by-side, how well their current product and the competition's are meeting customer needs (Diagram 6-12).

Any survey that asks customers how satisfied they are with your product can also be sent to your competition's customers. Most companies have ready access to their customers (or can get such access through their distributors). Access to the competition's customers may require more resourcefulness, such as making use of commercially available mailing lists, or surveying people at street corners, shopping malls, or trade conventions. Most market research firms can provide effective advice in this regard.

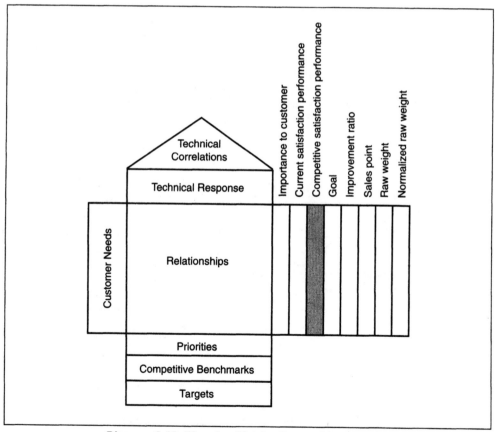

Diagram 6-12. Planning Matrix, Competition's Rating

As for benchmarking the competition's technical performance, this topic is generally beyond the scope of this book, although Chapter 10 covers this topic in a general fashion.

This allows them to set customer performance goals strategically. The team can choose to aim for high customer performance where the competition is weak, or for high customer performance where the competition is strong. By comparing Importance to Customer, Customer Satisfaction Performance, and Competitive Satisfaction Performance, a number of possible strategic choices become apparent. In Diagram 6-13, the comparison of Our Current Rating and Competition's Rating shows at a glance some important gaps.

"Controls under my fingertips" and "Enables me to find things in the document quickly" are areas in which our development team's product is substantially behind the competition. "Can customize to suit my working style" and "Offers lots of size, font, and design options" are areas where the development team is doing significantly better.

An alternate method of displaying ratings is by graphics. Many House of Quality diagrams in the literature show performance levels as points on a graph, connected by lines. The points may be represented by circles, squares, or triangles to distinguish the developer's product or service from the competition's. Diagram 6-14 illustrates this representation, using the same performance values that appear in Diagram 6-13. Most QFD software provides the

Customer Needs	Importance to customer	Customer satisfaction performance	Competitive satisfaction performance	Goal
Intuitive controls	84	2.9	2.8	
Controls under my fingertips	83	3.1	4.4	
Can customize to suit my working style	81	4.6	3.8	
Easy to get the information I need	80	4.7	4.6	
Can adjust the cursor to move as quickly as I'd like	49	2.9	2.8	
Enables me to find things in the document quickly	48	3.1	4.4	
Offers lots of size, font, and design options	45	4.6	3.8	
Able to see what the fonts look like as I'm choosing them	42	4.7	4.6	

Diagram 6-13. Planning Matrix, Strategic Comparisons

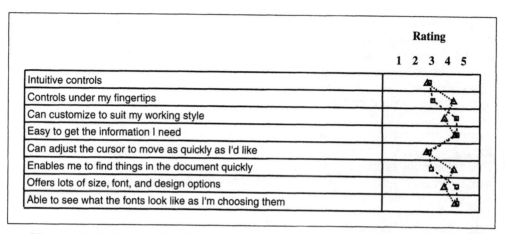

Diagram 6-14. Planning Matrix: Graphic Representation of Customer Performance

option for either representation. This author finds graphical representation, as in Diagram 6-14, difficult to read.

I worked with one team who attempted to fill in the customer performance section of the planning matrix only to discover that they knew almost nothing about their competition. The team guessed at the competition's strength, and where they could not agree, they entered question marks. Their planning matrix looked like Diagram 6-15.

Customer Needs	Importance to customer	Customer satisfaction performance	Competitive satisfaction performance	Goal
Intuitive controls	84	2.9	?	
Controls under my fingertips	83	3.1	5	
Can customize to suit my working style	81	4.6	?	
Easy to get the information I need	80	4.7	5	
Can adjust the cursor to move as quickly as I'd like	49	2.9	?	
Enables me to find things in the document quickly	48	3.1	?	
Offers lots of size, font, and design options	45	4.6	?	
Able to see what the fonts look like as I'm choosing them	42	4.7	3	

Diagram 6-15. Planning Matrix, Unknown Competitive Performance Levels

When their management discovered all the question marks instead of concrete data on the competition, they made a decision to launch a competitive bench-marking project. One of the benefits of QFD is that it creates a structure that in effect asks the right product planning questions. In this case the questions that still needed answers were painfully obvious, and the team knew what it had to do before it could complete its planning.

Competitive Performance data should appear in a QFD project in the same form as Customer Performance data. In the most typical marketing research scenario, surveys are designed to capture Competitive Performance in the same way as Customer Performance, so that the resulting quantitative data will be comparable.

Often, however, Competitive Performance data is not as neatly wrapped as we would like it to be. Most QFD teams "do the best they can" with such data. My suggestion in such a case is to gather together all the available Competitive Performance data, and present it in a single document. Sometimes the affinity diagram method is very helpful for sorting and organizing this data.

Having gathered Competitive Satisfaction Performance data, which may not have been collected in a form that matches the tree diagram of customer needs, the next step is to relate it to the customer needs. In cases where the data clearly lines up and fits the structured customer needs already established for the QFD, simply enter the data. There will probably be gaps where the Competitive Customer Performance data is not available. This should be indicated in the Planning Matrix.

One may be tempted to create a market research project to fill in all gaps of Competitive Performance. This is probably not necessary. I recommend getting new data only for those customer needs that emerge as very high priority after a first pass over the Planning Matrix.

6.4   Goal and Improvement Ratio

In the Goal column of the Planning Matrix (Diagram 6-16), the team decides what level of customer performance they want to aim for in meeting each customer need—the Goal. The performance goals are normally expressed in the same numerical scale as performance levels. The Goal, combined with Our Current Rating, is used to set the Improvement Ratio. The Improvement Ratio is one of the most important multipliers of Importance to Customer; thus, setting the Goal is a crucial strategic step in QFD.

Often, the question is asked: "Why set goals at all, or why not set all goals as high as possible—don't we want to excel in all areas?" This question relates

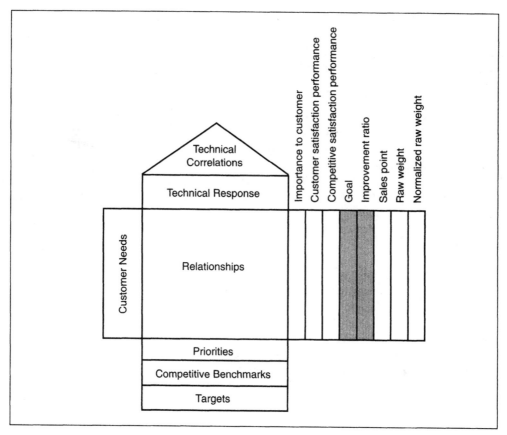

Diagram 6-16. Planning Matrix: Goal and Improvement Ratio

to one of the fundamental benefits of QFD. If we had unlimited resources, we could indeed aim to have our product or service be perfect in all aspects. But no projects ever *do* have unlimited resources. Development teams must always make choices regarding where we will place special emphasis or extra resources, and where we won't. Trade-offs must be made on all projects.

From the point of view of limited resources, it is a strategic necessity to choose which aspects of a product or service will excel, and which won't. Thus, goal setting in QFD involves comparing ourselves to the competition, and noticing which customer needs are most important. Setting performance goals in the planning matrix of the House of Quality generally has far-reaching effects on priorities throughout the development project.

This is because the goals, combined with Our Current Rating, determine the Improvement Ratio column, a measure of effort required to alter customer satisfaction performance for a customer attribute. If the company has a product or service in place, with a customer performance level of 3, say, for one customer attribute, then it will normally take about the same effort to achieve that same level of performance in a new version of the product or service as it did previously.

If the goal is higher than the current level—for example, 4, as compared to 3 for the previous product or service—then one may infer that "something special" will have to be done to have a positive effect on customer performance. "Something special" could mean an innovative redesign of at least part of the product or service, or it could mean a radical change in the way the product or service is packaged or delivered. Such changes often require corresponding multidepartmental changes. These types of changes are never easy to accomplish; thus, the development team must not take goal setting for customer performance lightly.

The arithmetic in the planning matrix is set up to reflect the difficulties of these changes, although the traditional QFD arithmetic as described in early QFD articles and books is not perfect. "Current Satisfaction Performance" and "Goal" are combined arithmetically to produce a value called the "Improvement Ratio." The Improvement Ratio is a multiplication factor which effectively scales the "Importance to Customer" and thus reorders the importance of the Customer Needs. The most common method for determining the Improvement Ratio is to divide the Goal by Current Satisfaction Performance:

$$\frac{\text{Goal}}{\text{Current Satisfaction Performance}} = \text{Improvement Ratio}$$

The more aggressive the Goal compared to Current Satisfaction Performance, the larger the Improvement Ratio, and hence the more important the Customer Need. However, this simple ratio may not provide the appropriate Improvement Ratio for many cases.

For instance, when Current Satisfaction Performance is very low (1, 2, or 3)—in other words, when the denominator of the Improvement Ratio is very low—then the Improvement Ratio itself will be quite large, even for modest improvement goals. On the other hand, when Current Satisfaction Performance is high (4 or 5), and therefore the denominator of the Improvement Ratio is high, then the Improvement Ratio itself will not influence the overall importance of the Customer Need very much. Let's look at a few examples (Diagram 6-17).

In cases 1, 2, and 3, the Goal has been set at two points better than Current Satisfaction Performance. The Improvement Ratios, however, range from 3 down to 1.67, differing by almost a factor of 2. This difference may or may not reflect the relative difficulty of achieving new levels of customer performance.

Case	Current Satisfaction Performance	Goal	Improvement Ratio
1	1	3	3
2	2	4	2
3	3	5	1.67
4	1	2	2
5	4	5	1.25

Diagram 6-17. Improvement Ratios

When customer performance is quite low, conventional quality wisdom suggests that it is usually relatively easy to make modest improvements. This same wisdom suggests that it becomes increasingly difficult to approach perfection. For example, in semiconductor manufacturing, yields of new products are traditionally quite low, perhaps as low as 20 percent. The production problems are generally fairly obvious, and yields can be brought up to 70 or 80 percent without great difficulty. As yields approach perfection, improvements become increasingly difficult.

This phenomenon is sometimes called the "Low-Hanging Fruit" factor. In the beginning of most quality improvement activities, many problems affecting quality are apparent to everyone and can be identified and dealt with easily. These correspond to the low-hanging fruit in the orchard: easy to see and easy to pluck. Once all the low-hanging fruit has been removed, the fruit higher in the tree remains. This fruit is harder to see, and when spotted is harder to pluck, because it cannot be reached by an orchard worker standing on the ground.

In the same way, when customer performance is low, problems often abound that are easily identified and easily fixed. When these obvious problems are out of the way, customer performance will have improved but still may not measure up to world-class competition. More sophisticated problem analysis and more elusive solutions may be required.

Referring back to the Improvement Ratio, one may now infer that it is easier to move customer performance from a 1 to a 2 than it is to move it from a 4 to a 5. Yet the Improvement Ratio arithmetic implies just the opposite: The Improvement Ratio of 5/4 is much smaller than the Improvement Ratio of 2/1. There are two alternative arithmetic approaches to the Improvement Ratio that can better reflect the Low-Hanging Fruit phenomenon. The first is to substitute an "Improvement Difference" for the Improvement Ratio, and the second is to use a "Degree of Difficulty" judgment directly.

The "Improvement Difference" is defined as

Improvement Difference = 1 + (Goal − Current Satisfaction Performance)

This formula has the characteristic that all improvement increments—whether starting from a low or a high level of customer performance—have the same impact on overall importance (raw weight) of a customer attribute. If the Goal is the same as Current Satisfaction Performance, a very common case, the difference would be zero, which obviously should not multiply the Importance to Customer. Hence, the Improvement Difference formula includes the "1+" term.

The formula has two disadvantages. First, in the rare case that the Goal is *less than* the Current Satisfaction Performance, the Improvement Difference will be negative or zero, thus making it an inconvenient multiplier of Importance to Customer. Second, the Low-Hanging Fruit theory suggests that it is much more difficult to improve customer performance when it is already high, as compared to when it is low. Thus, an even more realistic formula would express this sense of increasing difficulty.

The most straightforward way of dealing with the problem is to sidestep the formula completely, and directly specify the degree of difficulty:

1   no change
1.2   moderately difficult improvement
1.5   difficult improvement

Some QFD teams use Current Satisfaction Performance and Competition's Rating as reference values. They omit the step of setting an explicit Goal and enter one of these values into the Improvement Ratio column directly.

6.5   Sales Point

The Sales Point column (Diagram 6-18) contains information characterizing the ability to sell the product or service, based on how well each customer need is met. For example, for an automobile, a customer need might be for fuel efficiency. If the automobile could be designed to meet this need well, efforts to sell the product could capitalize on this capability.

The most common values assigned for Sales Points are

1   No sales point
1.2   Medium sales point
1.5   Strong sales point

Sales Points do not carry as much weight as other factors in the Planning Matrix, such as Importance to Customer or Satisfaction Performance Goal.

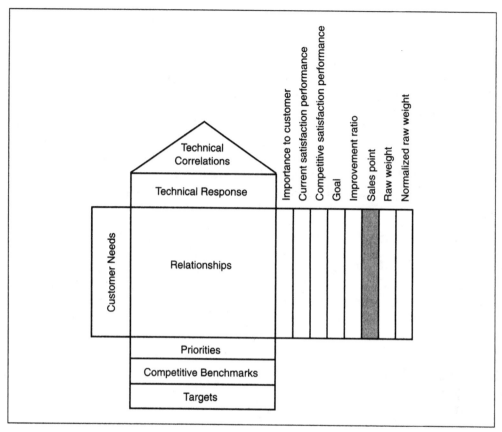

Diagram 6-18. Planning Matrix: Sales Points

This is because in the version of QFD that came from Japan, the ability to sell a product was not considered to be as important as the ability to increase customer satisfaction.

Obviously, performing very well on a customer need can also make it easier to sell the product. In this case, the Importance to Customer value might be high, the Goal value might be high, *and* the Sales Point value might be high. Some experts argue that a form of "double accounting" may be occurring, since the same general advantages are being expressed in all three values. This author acknowledges the possibility of such double accounting, but is not unduly troubled by it. The only harm in the possible double accounting is that the customer need's priority (raw weight) may be too high. With all these values high, performing well on this particular customer need should clearly be a priority in any case. QFD is not "rocket science"; the results should not be taken literally and blindly. The numerical manipulations give us a general idea of what's important, but must always be interpreted with common sense.

Not all customer needs represent sales opportunities. For example, fulfilling a need for safety or for compliance with long-established regulatory standards would not likely create customer interest that would justify a sales campaign.

Electronic devices such as hi-fi receivers always conform to Underwriters Laboratories (UL) standards in the United States. These standards aim at eliminating the possibility of electric shocks to users during normal operation.

It would be pointless, even counterproductive, for a company to advertise that their hi-fi receiver is shockproof. First, this safety characteristic is shared by all similar products, so it is not a differentiator. Second, the suggestion that products already *assumed* to be safe are indeed safe is likely to raise questions in the minds of potential customers about the safety claims.

In general, products whose characteristics meet "Expected" needs (in the sense of the Kano model or the Klein model [Section 16.2]) such as "shockproof" are not likely to be candidates for high values in the Sales Point column. Products or services that meet needs not met by the competition or by previous offerings, or that meet needs better than the competition or previous offerings, *are* candidates for high values in the Sales Point column.

Thus, strong sales points might be
- High fuel efficiency (automobile)
- Long life (light bulb)
- No need for a second coat (house paint)
- Gets clothes white and bright (laundry detergent)

How strong these sales points are depends on how they compare to the competition, and on how important it might be to the customer for the product to perform exceptionally well on these attributes.

At the time the Sales Point column is being filled in, the development team may have no idea what their design will be, or how they will meet specific customer needs. One way of harnessing QFD's power is to set aggressive goals in the Goal column of the Planning Matrix in customer need areas which could lead to competitive advantage, and then link the corresponding Sales Point values to those aggressive goals. This allows the QFD process to point out what parts of the design require breakthrough thinking in order to realize the advantage.

Guidance in deciding where to be aggressive can come from Kano Analysis (Section 2.8) or Klein Grid Analysis (Section 16.2). The Klein Grid model can help identify customer needs ("Hidden") which, if strongly met, could create disproportionate levels of customer satisfaction. The Kano model can help identify technical responses ("Delighters") to these Hidden needs.

6.6   Raw Weight

The Raw Weight column (Diagram 6-19) contains a computed value from the data and decisions made in Planning Matrix columns to the left. It models the *overall importance to the development team* of each Customer Need, based on its Importance to the Customer, the Improvement Ratio set by the development team, and the Sales Point value determined by the development team. The value of the Raw Weight for each Customer Need is

Raw Weight = (Importance to Customer) • (Improvement Ratio) • (Sales Point)

Using the most conventional formula for the Improvement Ratio,

$$\text{Improvement Ratio} = \frac{\text{Goal}}{\text{Customer Satisfaction Performance}}$$

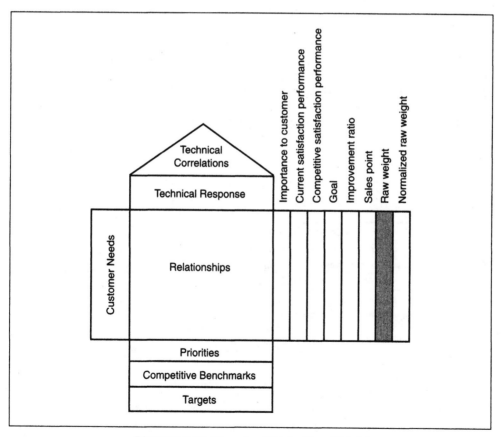

Diagram 6-19. Planning Matrix: Raw Weight

the Raw Weight formula becomes

$$\text{Raw Weight} = (\text{Importance to Customer}) \cdot \left(\frac{\text{Goal}}{\text{Customer Satisfaction Performance}} \right) \cdot (\text{Sales Point})$$

The higher the Raw Weight is, the more important the corresponding Customer Need is to the development team. The Raw Weight is a single number embodying customer satisfaction performance, implementation effort, and sales potential. Hence, it provides an overall strategic business perspective on the importance of the Customer Needs to the success of the product or service being planned.

One of the attractions of QFD to many development teams is that the Planning Matrix, along with the Raw Weight, provides a mechanism for funneling the priorities and concerns of interested parties across the organization into an analysis process that takes all concerns into account. It also helps the team to attach weights to the various concerns. These weights are determined by the value ranges used for each of the terms in the Raw Weight formula.

Notice that with the ranges of factors as given in Diagram 6-20, the Sales Point factor does not influence the Raw Weight as much as other factors, such as Importance to Customer. This reflects the philosophy that customer performance should be treated as more important than sales potential. Any QFD team is, of course, free to adjust the ranges of any of the factors to reflect the team's attitudes about which factors should most affect the planning process.

Factor	Minimum Value	Maximum Value
Importance to Customer (Absolute)	1	5
Importance to Customer (Weighted)	1	100
Customer Satisfaction Performance	1	5
Competitive Satisfaction Performance	1	5
Goal	1	5
Improvement Ratio (assuming $\text{Improvement Ratio} = \frac{\text{Goal}}{\text{Our Current Rating}}$)	0.2	5
Sales Point	1	1.5
Raw Weight (with Absolute Importances)	0.2	37.5
Raw Weight (with Relative Importances)	0.2	750

Diagram 6-20. Weight Ranges in the Planning Matrix

6.7 Normalized Raw Weight

The Normalized Raw Weight column (Diagram 6-21) contains the Raw Weight values, scaled to the range from 0 to 1 or expressed as a percentage.

To calculate the Normalized Raw Weight, first sum the Raw Weights to compute the Raw Weight Total:

$$\text{Raw Weight Total} = \sum \text{Raw Weight}$$

The Normalized Raw Weight for each Customer Need is then the Raw Weight for the Customer Need, divided by the Raw Weight Total:

$$\text{Normalized Raw Weight} = \frac{\text{Raw Weight}}{\text{Raw Weight Total}}$$

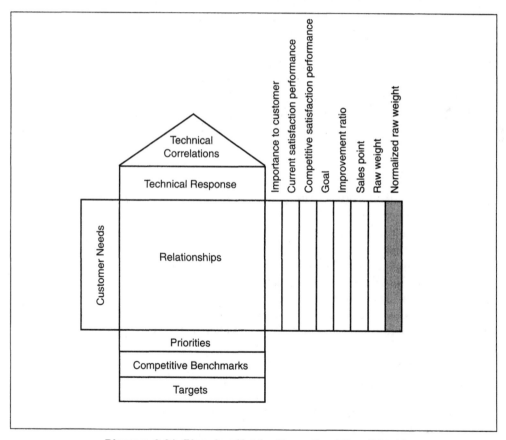

Diagram 6-21. Planning Matrix: Normalized Raw Weight

The Normalized Raw Weight will be a fraction between 0 and 1. Some people prefer expressing the Normalized Raw Weight as a percentage, rather than as a fraction. To express it as a percentage, multiply the Normalized Raw Weight by 100. As a fraction or as a percentage, the values are equivalent, and we won't distinguish between them in this book. In Diagram 6-23, the worked-out example of the Planning Matrix, the Normalized Raw Weights are expressed as fractions.

Since the Raw Weight will be used in QFD as a proportional value, the Normalized Raw Weight carries the same information as the Raw Weight. In other words, if the Raw Weight for Customer Need A is double the Raw Weight for Customer Need B, the same ratio will apply for the Normalized Raw Weights of these two Customer Needs.

It's convenient to convert the Raw Weight to a Normalized Raw Weight for subsequent calculations in QFD. The Raw Weight, which is often in the range of 15 or higher, will be multiplied by other values in the Relationships section of the HOQ (see Chapter 8). These multiplied values will be added, and the resultant sums can be above 1000. If the Normalized Raw Weight is used instead, the resultant sums will be much smaller, and therefore generally easier to manage and display. As we have seen elsewhere, these sums will used as weights to be transferred to other matrices, where they will in turn be multiplied by other numbers and added to create a new set of weights. These weights tend to get large, and most QFD practitioners reduce them again by converting them to Normalized Weights as described here.

6.8   Cumulative Normalized Raw Weight

The Cumulative Normalized Raw Weight, when used, is normally placed last, to the extreme right of the Planning Matrix, as in Diagram 6-22. After the Raw Weights and Normalized Raw Weights have been computed, and therefore the Customer Needs have been prioritized, it is sometimes useful to view the Customer Need raw weights in terms of their overall importance. To do this, the team sorts the Customer Needs by Normalized Raw Weight in descending order (with the highest weight first), as in Diagram 6-23.

Once the customer needs have been rearranged this way, it's possible to identify the most important customer needs as a group. The Cumulative Normalized Raw Weight column is based on the values in the Normalized Raw Weight column. The Cumulative Normalized Raw Weight shows how much of the total raw weight can be attributed to the most important customer need, the two most important customer needs, the three most important, and so on.

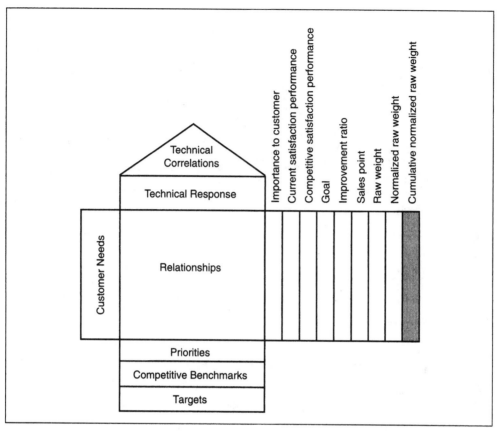

Diagram 6-22. Planning Matrix, Cumulative Normalized Raw Weight

Each Cumulative Normalized Raw Weight value in a row is formed from the sum of the Normalized Raw Weight of the customer need of the row, and the Normalized Raw Weights of all the customer needs more important (higher) than this need. By displaying the cumulative weights this way, we can easily see which customer needs contribute to various fractions of the total raw weight. For example, the most important needs, which as a group contribute to one-half the total raw weight, are found by scanning the Cumulative Normalized Raw Weight column from the top downward, to the first value that is equal to or greater than 0.50. In Diagram 6-23, we see that the first three customer needs contribute one-half the total Raw Weight. Frequently a small subset of customer needs accounts for most of the raw weight. This information can be useful in saving time during the QFD process (see Section 15.7.2), and, more importantly, in focusing the team on the most important project goals.

	Weighted Importance to Customer	Customer Satisfaction Performance	Competitive Satisfaction Performance	Goal	Improvement Ratio	Sales Point	Raw Weight	Normalized Raw Weight	Cumulative Normalized Raw Weight
Can customize to suit my working style	81	4.6	3.8	4.6	1.00	1.5	559	0.19	0.19
Easy to get the information I need	80	4.7	4.6	4.7	1.00	1.2	451	0.16	0.35
Controls under my fingertips	83	3.1	4.4	4.4	1.42	1.2	438	0.15	0.50
Intuitive controls	84	2.9	2.8	3.3	1.14	1.5	416	0.14	0.65
Enables me to find things in the document quickly	48	3.1	4.4	4.5	1.45	1.5	324	0.11	0.76
Offers lots of size, font, and design options	45	4.6	3.8	4.6	1.00	1.5	311	0.11	0.87
Able to see what the fonts look like as I'm choosing them	42	4.7	4.6	4.7	1.00	1.2	237	0.08	0.95
Can adjust the cursor to move as quickly as I'd like	49	2.9	2.8	2.9	1.00	1.0	142	0.05	1.00
Totals							2878	1.00	

Diagram 6-23. Completed Planning Matrix

Summary

The Planning Matrix is the portion of the House of Quality that contains strategic marketing information and planning decisions. Completing this section of the House of Quality is a major step in the QFD process. It creates a set of priorities for the customer needs that are based on their importance to the customer *and* their importance to the development team's organization. This set of priorities will have a major impact on all future planning and development activities.

The quantitative information in the Planning Matrix normally is the result of market research. The team makes its strategic decisions based on their understanding of the customer needs, and their own and their competition's performance in meeting those needs. The most important and far-reaching strategic judgments are the setting of customer satisfaction performance goals and the identification of selling opportunities for the product or service being planned.

The information that feeds the decision making in the Planning Matrix is

- Importance to the customers of each customer attribute
- Current Customer Satisfaction Performance, by Customer Need, with the development team's product or service most similar to the one being planned
- Current Customer Satisfaction Performance, by Customer Need, with the competitor's product or service most similar to the one being planned

The strategic determinations the development team makes and records in the Planning Matrix are

- Customer Satisfaction Performance Goal for each customer attribute
- The sales potential of performance on each customer attribute

The computations the development team makes in the Planning Matrix are

$$\text{Improvement Ratio} = \frac{\text{Goal}}{\text{Customer Satisfaction Performance}}$$

$$\text{Raw Weight} = (\text{Importance to Customer}) \cdot (\text{Improvement Ratio}) \cdot (\text{Sales Point})$$

and

$$\text{Normalized Raw Weight} = \frac{\text{Raw Weight}}{\text{Raw Weight Total}}$$

The Raw Weight and the Normalized Raw Weight convey equivalent information, but the Normalized Raw Weight is more convenient in later phases of QFD. Either calculation provides a rank ordering of customer attributes which will influence decision making throughout the planning process.

We've now covered QFD's handling of the Voice of the Customer both qualitatively and quantitatively. It's time now to consider the organization's response to these customer needs. Up to now we've listened to the customer speaking the customer's own language. In the next chapter, we'll look at the organization's language, and how QFD helps the organization speak its language.

Discussion Questions

In your organization, how do you determine the importance to the customer of the various customer attributes? What do you like about the method? How could it be improved?

How do you set customer satisfaction performance goals? Do you know customer performance levels for each customer attribute? If not, how could you gather such information? Do you know how well the competition is doing?

What strategic planning decisions do your developers make, and how do they compare to the decisions identified by the columns of the Planning Matrix? If you wanted to record your organization's planning decisions in a structure like the Planning Matrix, which columns would you add, delete, or change?

---

[1] Glen Urban and John Hauser, *Design and Marketing of New Products,* 2nd edition, Prentice Hall, Englewood Cliffs, New Jersey, 1993.

CHAPTER 7

Substitute Quality Characteristics (Technical Response)

Introduction

This chapter describes the way in which QFD deals with the development team's technical response to the customer's needs. Just as the Voice of the Customer had a qualitative and quantitative component (entered into the Customer Needs/Benefits Section and the Planning Matrix), so does the translation of the Voice of the Customer into the Voice of the Developer. The translation, which we'll usually call the Substitute Quality Characteristics (SQCs), will be placed in qualitative form on the top of the Relationships Matrix, and in quantitative form at the bottom (Target Values and Competitive Technical Benchmarks).

In this chapter we'll look at a few alternative formulations for the SQCs, along with some methods for generating them.

Substitute Quality Characteristics is the term used for the internal, technical language an organization uses to describe its product or service. In QFD parlance, we use the term Quality Characteristics to denote the customer's needs (the Voice of the Customer). The translation into technical terms is called Substitute Quality Characteristics because it represents a translation from the customer's language into the organization's technical language. This Technical

Response (Diagram 1-1, Section C, and Diagram 7-1) may describe the product from any of a variety of points of view. Most commonly, developers call this language the Product Requirements or the Design Requirements.

The nature of Product Requirements varies widely from group to group, and from industry to industry. Many organizations describe products and services in more than one language, or more than one set of terms. They distinguish between these various languages by giving them such names as Customer Requirements, Market Requirements, Top-Level Specifications, Detailed Specifications, or Technical Specifications, to name but a few.

Some generic formulations of Substitute Quality Characteristics exist, notably one by Stuart Pugh (see Section 12.2, Enhanced QFD). Such formulations can be used as "starter kits" to get a set of SQCs established rapidly, and to aid teams in arriving at a complete set. The generic formulations must of course

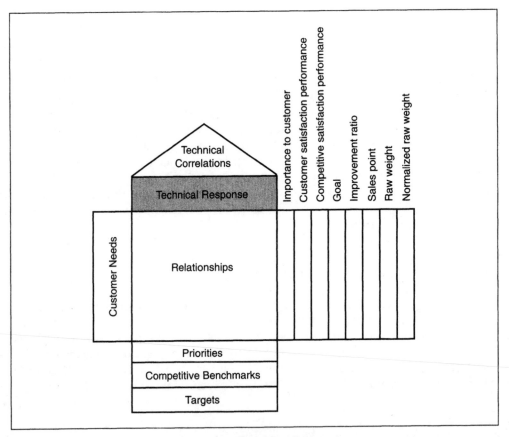

Diagram 7-1. Technical Response

be modified to meet the needs of the team's specific project. In particular, the language of the generic formulations is not likely to correspond to the development team's product description languages.

There is often much confusion about the boundaries between these various product description languages, and there is often little or no standardization of the vocabulary of each language. In very hierarchical organizations, the most workable distinction between Top-Level Specifications and Detailed Specifications, for example, may be defined by what organization is responsible for writing the associated documents.

The ideal relationship between various product or service description languages is one in which the languages are defined and ordered according to how abstract, or solution-independent, each language is. If one product description allows for many possible implementations, it is more abstract than another which clearly describes or implies one and only one implementation.

In software, for example, a product requirement for "easy command selection" is more abstract than "commands available by pull-down menus." "Easy command selection" allows for at least the three most popular technical solutions—pull-down menus, iconic push-buttons, and keyboard shortcuts. "Commands available by pull-down menus" only allows for one of these possibilities and is therefore more concrete, or less abstract.

Any development organization would do well to define its highest-level product or service description language by providing a vocabulary of terms that are included in the language, along with examples. Lower-level languages could be similarly defined. Most important to these definitions would be to show how a designer translates a statement from a higher-level, more abstract language to one or more statements in a lower-level, more concrete language.

One way that QFD practitioners describe these various levels of abstraction is with the "Whats/Hows" metaphor (Diagram 7-2). The language that appears on the left side of a QFD matrix represents "What" is desired. The language at the top of the matrix describes "How" the developers will respond to the "Whats." To get the most out of QFD, the language of the "Whats" should be distinctly more abstract than the language of the "Hows."

The "Hows" may still be abstract, compared to other languages available to the developers and to be used later in the development process. In later phases of development, and later phases of QFD, the "Hows" can be placed on the left side of another matrix so that they become the "Whats" at a more detailed level (Diagram 7-3). The development team would then use more specific language along the top of the new matrix to represent the "Hows" of that matrix.

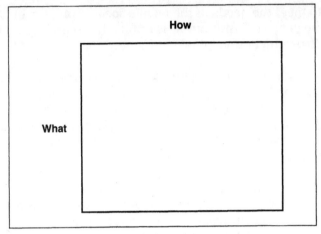

Diagram 7-2. What versus How

Given a term that describes some aspect of a product, it is not always easy to decide what language it belongs to. An example given elsewhere in this book (see Chapter 16) illustrates the common situation of a customer expressing a need as a solution. The customer asks for "tinted glass" in an automobile instead of asking for "privacy." Because "tinted glass" is technically the Voice of the Customer, shouldn't we include "tinted glass" with the "Whats" and place it on the left side of the House of Quality? On the other hand, because "tinted glass" is a technical response to the customer need for "privacy," shouldn't we place it along the top of the House of Quality?

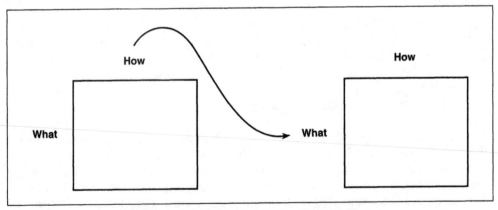

Diagram 7-3. Deploying "Hows" to "Whats"

A practical method for dealing with the problem of placing product requirements in the right place in QFD is to use the Voice of the Customer Table
(VOCT) described in Chapters 5 and 16. Even if the development team lacks
preexisting definitions for their technical description language(s), the VOCT
provides a concrete method for sorting out terms and for deciding how many
levels of technical description languages they want to work with.

7.1   Top-Level Performance Measurements

Probably the most valuable language for Substitute Quality Characteristics is
the language of Performance Measurements. These are measurements that the
development team derives directly from customer needs. They should be general enough to be applied to a product or service regardless of the specific
implementation. Therefore, they provide ideal measurements for benchmarking competitive products or services and for providing a solution-independent
starting point for developing new concepts.

The standard method for developing Performance Measurements is to begin
with the customer attributes. For each customer attribute,

1. Define measures
2. Define measurements

7.1.1   Define Measures

Defining measures is the process by which the development team establishes
the relevance and the relationship between its measurements and customers'
perceptions. The word "Deployment" in "Quality Function Deployment"
most readily applies to this process of defining measures. In a nutshell, the
team translates, or deploys, each customer need into a technical performance
measure.

For each customer attribute, define one or a few technical performance measurements. Write these along the top of the House of Quality, as in Diagram 7-4.
Some examples of performance measures and their relationship to customer
needs are shown in Diagram 2-8.

For each measurement, be sure of the following:

- That it can be measured while the product or service is being developed,
  and before it is shipped or deployed; in other words, that it can be used
  as a predictor rather than a lagging indicator of customer satisfaction
  performance.

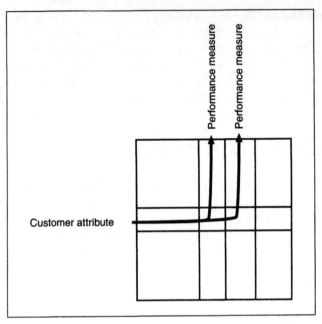

Diagram 7-4. Customer Attribute Deployed to Performance Measurement(s)

- That it can be controlled by the development team. The team should be able to make decisions that effectively would adjust the measurement up or down. A good way to think of the measurement is as a "knob" the development team can rotate clockwise or counterclockwise. As the team conceptually rotates the knob, customer satisfaction performance may conceptually be affected either positively or negatively.

To be properly defined, each performance measurement should be characterized in a few ways. First, the *units* of the measurement should be defined. Examples would be

- Voltage in volts
- Time in minutes
- Process complexity in number of steps
- Accuracy in defects per thousand transactions

Second, the *direction of goodness* should be defined. There are three possible directions of goodness:

- *The More the Better.* The implied target is infinity. Examples are the following:

    Reliability as measured by mean time between failure (MTBF).
    Fuel efficiency as measured by miles traveled per gallon of fuel used.

Bonding strength of an adhesive as measured by pounds supported per square inch of adhesive area.

The number of instances of treating a customer respectfully.

Some development teams confuse the implied target of infinity for *The More the Better* with an acceptable value (a *tolerance*) for the metric. Many developers show consternation when confronted with targets such as infinity. The idea behind using such language is to recognize that most technical objectives are points on a sliding scale. If we view them as such, we may see possibilities for increasing performance that we have not seen before.

On the other hand, arbitrarily high numerical objectives are usually impossible or very impractical to achieve. Teams need both the *aggressive* objective (infinity) and an *acceptable* value for the measure. While the market may *accept* a product with reliability of twenty thousand hours mean time between failure, clearly a product with much higher MTBF would be more attractive, if all other factors—including price—were the same.

- *The Less the Better.* The implied target is nominally zero. Examples are the following:

    Quality of service as measured by number of defects per transaction. Defects could be determined by monitoring all or a sample of transactions. If the transactions are interpersonal—for example, hot-line service center calls—the defects could be instances of treating a customer disrespectfully, or instances of giving a customer incorrect information.

    Simplicity of process as measured by process complexity in number of steps.

    Speed of startup as measured by time to launch a software application.

There may be rare cases where the target is minus infinity (as for certain refrigeration applications where "Target is best" does not apply), but this author has never seen such a case. Sometimes in a case where the target is minus infinity, the measurement scale can be shifted so that zero becomes the lowest possible value. In the refrigeration example, absolute zero could be selected as the ideal value.

- *Target Is Best.* The target is as close as possible to a nominal value with no variation around that value. Examples are the following:

    Exactness of fit as measured by the diameter of a steel rod intended to fit into a cylindrical sleeve. If the diameter is too large, the rod will not fit into the sleeve; if the diameter is too small, the rod will wobble in the sleeve.

    Constancy of ideal temperature within a food freezer container. The ideal temperature is 4°F. Colder temperatures cause the food's flavor to deteriorate; warmer temperatures reduce the shelf life of the food.

The majority of top-level performance measurements in service applications will be of the "More the Better" and "Less the Better" types. "Target Is Best" is not common for service applications.

7.1.2   Define Measurements

Describe *how* each measurement will be performed. Also, document all assumptions and comments about each measurement.

Describing how each measurement will be performed is a step that escapes many developers. They may feel that the measurement method is self-evident and doesn't need explicit description. The omission of this step leads to much lost time during planning and later on during development.

This step *operationalizes* the definition of the measurement. Deming points out that definition of the process of measurement is a key factor in defining the measure.[1] He says that operationally defined measures are measures "one can do business with." Conversely, measures that are not operationally defined are measures one *cannot* do business with. Such measures cause confusion, because one person will inevitably have in mind a different measurement process than another. Let's look at an example:

Consider the following measure: "Speed of startup as measured by time to launch a software application." One developer may assume the method of measurement involves

1. No other applications running on the computer.
2. Operating system is version 3.1, as shipped by the operating system supplier, installed with supplier standard defaults.
3. Maximum RAM configuration will be used for measurements.
4. Clock starts when the "Open application" command is executed at the operating system level.
5. Clock stops when the application is ready to receive a user command.
6. Launch five times and take the average of the five launches.

Another developer may have a very different set of assumptions:

1. Other applications are running on the computer, in particular applications A, B, C, and D, which many customers expect to use in conjunction with the application.
2. Launch under operating system versions 2.9, 3.0, and 3.1 must all be measured. Each operating system will be used as shipped by the operating system supplier and installed with supplier standard defaults.

3. Minimum and maximum RAM configurations will used for measurements.
4. Clock starts when the "Open application" command is executed at the operating system level.
5. Clock stops when the application is ready to receive a user command.
6. Identify sixteen configurations using different combinations of applications A through D, operating system version 2.9 through 3.1, and minimum and maximum RAM configurations. Launch once with each configuration and take the average of the sixteen launches.

Developers working to the first definition would optimize application launch for the single configuration specified. Such an application might do very poorly under minimum RAM, or with operating system 2.9. The detailed description of the measurement guides developers about what to optimize. It makes all discussions involving the measurement clearer and more efficient.

7.2   Product Functions

A completely different approach to defining Substitute Quality Characteristics is to place *product* or *process functions* along the top of the House of Quality. It can be appropriate to use functions instead of performance measures under the following circumstances:

- The product or service concept has already been established (breakthrough ideas, at least at the strategic level, are not needed or are already incorporated into the concept). Many times a successful version of a product or service is already in the field, and QFD is being used to define a "midlife kicker" or an upgrade to the previous offering. In such a case there may be a list of possible extensions—already expressed as features—that need to be prioritized.
- The development team lacks either the time or the interest to develop and prioritize performance measures. Since prioritization of performance measures does not define a product's or service's features, the QFD process must be used at least once more to translate prioritized performance measures into prioritized features. This extra step is time-consuming and may not always be worth the effort.

In addition, some development teams, especially software development teams, are unaccustomed to using performance measures in their product definition process. For such teams, these measures may only get in the way of what they see as their work. While translating customer needs directly into functions lowers the chances for breakthrough ideas, teams that don't normally use performance measures may be better doing just such a translation.

Many products and services have large numbers of capabilities or functions. It may be bewildering to place them all at the top of a House of Quality. Depending on the level of detail the developers use to describe the functions, the House of Quality could be correspondingly small or large.

The developers can use the Affinity Diagram process to decide what level of functional detail they want to work at. The Affinity Diagram hierarchy of product or process function will have several levels (as do all Affinity Diagrams). Analyzing at the higher levels (with fewer elements) will present the advantage of quicker analysis. The corresponding disadvantage of working with less detail is less depth in the analysis.

Generally, for strategic analysis, a quicker analysis phase at a less detailed level is appropriate. The House of Quality at the strategic analysis level will indicate which few critical functional areas require more detailed planning. These areas can then be singled out, and the development team can analyze only those areas in a subsequent House of Quality (see Diagram 7-5).

7.2.1   Function Trees

In the previous section we described a brainstorming approach for creating an Affinity Diagram of the functions of a product or service. As is true of all Affinity Diagrams, functions are organized from the bottom up.

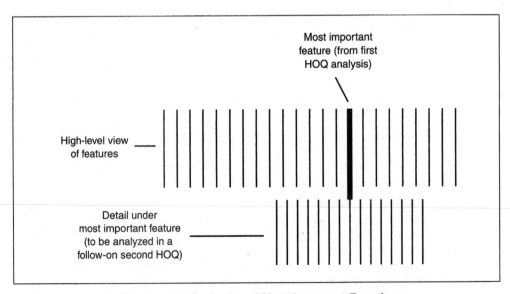

Diagram 7-5. Explosion of Most Important Function

Another approach, the Function Tree method, uses the Tree Diagram method and organizes the functions from the top down. This method has been described by Don Clausing in recent writings.[2]

In this approach, the primary functions of the product or service are identified first. Each primary function is then broken down into subfunctions. Each subfunction is elaborated into finer detail until the development team has reached the level of detail it needs. This top-down approach creates a *function tree* that helps the development team to focus on the most important functions of their product or service.

The resulting Tree Diagram can have multiple levels of increasing detail, just like an Affinity Diagram. The development team can choose the level on which to perform their QFD analysis.

A Function Tree may be indistinguishable from an Affinity Diagram of functions, once it has been completed. The method of creating it and the associated point of view are different—just as affinity diagramming and tree diagramming are different. To get the best of both methods, the team may consider first brainstorming functions and affinitizing them, then completing the structure with the Tree Diagram method by working from the top down.

7.3   Product Subsystems

While the design elements of a product are not normally chosen to be the Substitute Quality Characteristics, there are occasions where the choice is appropriate. Most commonly the QFD process translates from

1. the Voice of the Customer, to
2. Performance Measures, to
3. Functions, to
4. Product Design.

Each successive pair of topics (1 to 2, 2 to 3, 3 to 4) represents the left and top, respectively, of a new matrix. We have already discussed the possibility of a Voice of the Customer to Functions matrix (1 to 3). Such analysis would be appropriate if the product concept was "static." We'll be discussing static/dynamic analysis later on (Section 12.2.1).

Development teams have also created Voice of the Customer to Product Design matrices (1 to 4). Such a matrix shows the development team how various components of the product design influence various customer satisfaction attributes.

The most common method for representing the design is by use of the Tree Diagram. First, identify the primary subsystems of the product. For example, with a camera, the primary subsystems would be

> Imaging subsystem
> Film management subsystem
> Viewfinding subsystem
> Exposure time management subsystem

Every part of the camera should be contained within one and only one of these subsystems. The imaging subsystem might contain:

> Lens
> Film plane
> Lightproof compartment between lens and film

The film management subsystem might contain:

> Take-up spool
> Film advance subsystem
> Film plane subsystem (holds the film flat for accurate focusing)
> Film supply subsystem

Each of these subsystems and components can be described at several levels of additional detail, thus providing the development team with the multiple-level Tree Diagram it needs for QFD analysis.

7.4   Process Steps

For teams developing new processes or services, the following choices for Substitute Quality Characteristic are as applicable as for products:

- Performance measures
- Functions

Performance measures for services are much the same as for physical products: their values are generally under the control of the team that designs or lays out the processes underlying the service. Each process will have one or more clearly defined inputs and outputs. Development teams can define performance measures for these processes, based on time, cost, or quality of result. Examples of such measures are the following:

> Average time to process an input and produce the output (cycle time)
> Average number of inputs processed per unit time (throughput)
> Number of errors per standard number of transactions (quality; defect rate)
> Average processing cost per input (cost)

In QFD, the service development team can evaluate these and many other process performance measures to determine which ones drive customer satisfaction performance most strongly.

While product subsystem analysis may not apply to services, a closely related method of analyzing services and processes does apply. Services are delivered by processes. These processes may be conceived of at various levels of abstraction. For example, "Telephone Customer Support" for a financial institution, such as a bank, could be conceived of—at a very abstract level—as a simple process, as in Diagram 7-6:

1. Route incoming call to Customer Service Associate.
2. Classify customer request.
3. Respond to customer request.

Anyone who has analyzed telephone customer support will know that beneath this simple process lies an enormously complex structure for handling customer requests. This structure includes many decision points and subprocesses at each of the three main steps listed here. It also includes tools and technology to support these decision points and steps.

The decisions relate to managing the flow of incoming calls and balancing the volume of these calls against the available Service Associates; to identifying the customer, the customer account, and the type of customer request; to deciding on the appropriate response and acting accordingly. The tools and

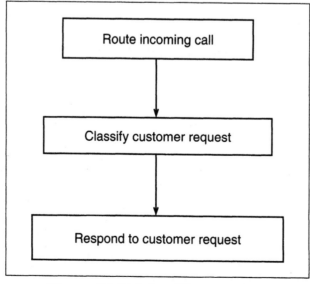

Diagram 7-6. High-Level View of Process

technology to support these steps include elaborate telephone equipment to queue and route incoming calls, as well as computers and associated databases to provide up-to-the-minute information about each customer to the Service Associates, and to update this information based on the nature of the customer call.

Each of these enumerated service components represents a subsystem or sub-service of the overall service. Each is a candidate to be a Substitute Quality Characteristic in the House of Quality. These service components can be evaluated for their impact on meeting customer needs.

The relative advantages and disadvantages of using process steps as compared to performance measures are shown in Diagram 7-7.

Summary

Substitute Quality Characteristics represent the organization's internal, technical language. The development team uses SQCs to describe its product or service in an abstract manner. The QFD team must choose which of its possibly many technical formulations it wishes to use in QFD. More abstract, solution-independent formulations provide the team with more breakthrough opportunities at the expense of more QFD steps to bridge the gap between customer wants and needs and action.

Defining a language for Substitute Quality Characteristics can be assisted by referring to the "Whats versus Hows" model. Learning to separate problems from solutions is an early benefit frequently cited by QFD teams.

The most commonly used Substitute Quality Characteristics are Top-Level Performance Measurements.

Performance measures	Process steps
Advantages: 1. Generally solution-free, providing stronger likelihood of creative solutions. 2. Measurements can be used to manage the processes.	Advantages: 1. Concrete, easily envisioned. 2. Can be used in the HOQ at the level of detail appropriate to the team's needs.
Disadvantages: 1. Difficult to understand or to create in organizations where measurement is not the norm. 2. Expensive to implement (cost—benefit issues aside).	Disadvantages: 1. Incomplete definitions of process steps can lead to confusion during QFD. 2. Focus on concrete process steps too early reduces the chances for breakthrough solutions.

Diagram 7-7. Performance Measures Versus Process Steps

Other SQCs include Product Functions, Product Subsystems, and Process Steps. In QFD applications where the SQCs are Top-Level Performance Measurements, the team generally develops a subsequent QFD matrix that puts the rank-ordered Top-Level Performance Measurements on the left, and one of these more specific forms on the top.

At this point, we know the Voice of the Customer and the Voice of the Developer. The next question that QFD asks, and that the development team must answer, is, how are these two voices linked up? If we developed the SQCs from the Customer Needs, we have confidence that each SQC is linked strongly to at least one customer need. But that may be the tip of the iceberg. There may be many other important linkages between SQCs and customer needs. We need to study them in order to decide which of the SQCs are powerful drivers of customer satisfaction. In the next chapter, we'll see how these linkages are evaluated.

Discussion Questions

In your development process, compare your description of Customer Wants and Needs with your description of your Technical Response. Can you separate the Whats from the Hows? Do your colleagues agree with your analysis?

What language do you use for your Technical Response? List some typical nouns and verbs. Is your language solution-independent? What would be a more solution-independent language? What would be a more solution-specific language?

---

[1] W. Edwards Deming, "Out of the Crisis," Massachusetts Institute of Technology, Center for Advanced Engineering Study, Cambridge, Mass 02139, Chapter 9. See also, W. Edwards Deming, "Quality, Productivity, and Competitive Position," Massachusetts Institute of Technology, Center for Advanced Engineering Study, Cambridge, Mass 02139, Chapter 15.
[2] *Total Quality Design*, Dr. Don Clausing, ASME Press, 1994.

CHAPTER 8

Impacts, Relationships, and Priorities

Introduction

This chapter shows how the relationships between Substitute Quality Characteristics and Customer Needs are modeled in QFD. One of the particularly brilliant ideas embedded in QFD is to use a matrix to study these relationships. In the past, the relationship between Technical Response and Customer Need was normally expressed by simple lists, and the complex many-to-many relationships remained only intuitively comprehended.

We will see how the Relationships matrix makes it possible to represent and visualize the patterns of these relationships much more easily than could ever be done before.

In this chapter we'll see how QFD handles strong and weak relationships, and even negative relationships.

At the point when the development team is ready to fill in the Relationships, the development team will have already determined a substantial amount of information about the marketplace. The Customer Needs will have been determined, the strategic market research information and decisions in the Planning Matrix will have usually been determined, and the Technical Responses or Substitute Quality Characteristics will have been decided upon.

The team's next task is to fill in the Relationships section. This chapter will show how the development team does this. We will also see how the team determines each SQC's relative contribution to overall customer satisfaction, and therefore its priority.

8.1   Amount of Impact

The Relationships section (Diagram 8-1) provides a mapping between Substitute Quality Characteristics on the one hand, and Customer Wants and Needs on the other.

Each Relationship cell represents a judgment, made by the development team, of the strength of the linkage between *one* Substitute Quality Characteristic

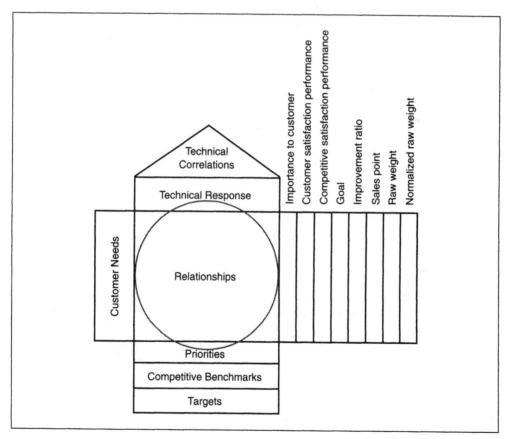

Diagram 8-1. Relationships

and *one* Customer Need. We call the strength of linkage the *impact* of the Substitute Quality Characteristic on the Customer Need. The entire Relationship Section of the HOQ contains cells for storing these impacts about *each* SQC/Customer Need pair.

We now come to the reason why the analogy of the Substitute Quality Characteristic as a "knob" is useful (see Section 7.1.1.) As the Substitute Quality Characteristic's "knob" is conceptually rotated in the direction of goodness (or in the opposite direction), what will happen to customer satisfaction performance with respect to a particular customer need? In QFD we recognize four possibilities:

1. Customer satisfaction performance with respect to the need is *not linked* to the SQC. In other words, for changes of any sort, large or small, in the amount or the degree of the SQC, no noticeable change in customer satisfaction performance of that need is predicted by the development team.
2. Customer satisfaction performance with respect to the need is *possibly linked* to the SQC. For relatively large changes in the amount of the SQC, little or no change in customer satisfaction performance of that need is predicted by the development team.
3. Customer satisfaction performance with respect to the need is *moderately linked* to the SQC. For relatively large changes in the amount of the SQC, noticeable but not major changes in customer satisfaction performance of that need are predicted by the development team.
4. Customer satisfaction performance with respect to the need is *strongly linked* to the SQC. For relatively small changes in the amount of the SQC, significant changes in customer satisfaction performance on that need are predicted by the development team.

For most QFD activities, the linkage is considered to be positive; that is, if the SQC is moved in the direction of goodness, customer satisfaction is assumed to increase. Negative linkage is possible, but it complicates the QFD process. Whenever possible, the development team should try to convert such negative linkage to positive linkage by redefining the SQC. Unfortunately, this is not always possible. Section 8.4 provides a detailed discussion of the special considerations for handling negative linkage.

Certain symbols are customarily used in QFD to denote these four possible impacts. The symbols, their meanings, and their numerical equivalents are as shown in Diagram 8-2.

The symbol meaning "Not linked" is a blank. Occasionally this causes a bit of confusion in QFD, because it is not possible to distinguish a matrix cell that has been evaluated as "Not linked" from a cell that has not been evaluated.

Symbol	Meaning	Most Common Numerical Value	Other Values
	Not linked	0	
△	Possibly linked	1	
○	Moderately linked	3	
◎	Strongly linked	9	10, 7, 5

Diagram 8-2. Impact Symbols

Some QFD facilitators use a symbol such as a check mark or a lower-case "b" in place of the blank in order to avoid this confusion.

In Diagram 8-3, one can see at a glance that Technical Response X makes a much greater contribution to customer satisfaction performance than does Technical Response Y. The diagram also suggests that Need A is being addressed more directly than Need B: Need A has a ◎ and a △ to its credit, while Need B has only a single ○.

8.1.1   Judging Impact of Performance Measures

The Performance Measure is the ideal Substitute Quality Characteristic for making impact judgments. The Performance Measure can be thought of as a continuous variable, and customer satisfaction performance for any need is normally represented as a continuous variable (a simplification of the statistical

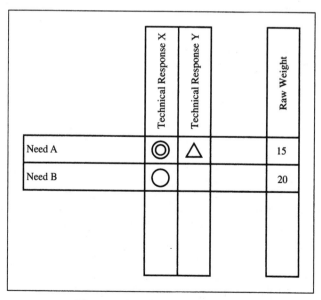

Diagram 8-3. Amount Of Impact

basis for measuring satisfaction performance). Thus, the impact represents the mathematical relationship between the two variables:

Customer Satisfaction $_A = f$(Performance Measure $_X$)

In this context, the impacts ◎, ○, and △ correspond to steep slope, moderate slope, and almost no slope. A blank impact corresponds to no slope.

This discussion assumes a *monotonic* relationship between Performance Measure $_X$ and Customer Satisfaction $_A$. In other words, as the Performance Measure moves in the direction of goodness, customer satisfaction performance continues to improve. Engineers are aware that the relationships may be more complex than that. To make product planning simpler and cleaner, it's a good idea to try to define Performance Measures that do, in fact, provide a monotonic relationship with customer satisfaction performance.

In Diagram 8-4, not only have we modeled the relationships as monotonic, we have modeled them as *linear*. In other words, as the Performance Measure moves in the direction of goodness, customer satisfaction performance continues to improve *at the same rate*. As we have seen with the Kano model, however, the relationships for Delighters and for Dissatisfiers are not linear (see Diagram 2-6). Delighters have the potential of disproportionately increasing customer satisfaction as the SQC moves in the direction of goodness, while Dissatisfiers have the potential of producing disproportionately increasing customer dissatisfaction as the SQC moves in away from the direction of goodness.

This type of disproportionate, nonlinear behavior cannot be easily modeled in QFD. The way to handle it is to treat such SQCs as if they were linear, but to

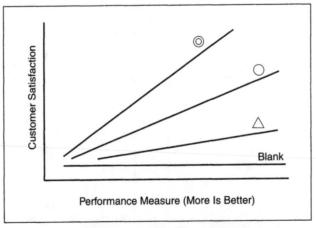

**Diagram 8-4. Customer Satisfaction Performance
as a Function of a Performance Measure**

note their special characteristics when the time comes to set target values. We'll be discussing target value setting in Chapter 11.

8.1.2   Judging Impact of Other SQCs

It is generally more difficult for development teams to judge impacts when the Substitute Quality Characteristics are not measurable. This is because a simple diagram such as Diagram 8-4 cannot readily be used as a conceptual model of the relationships. There is no clear and simple "knob" to turn, when the SQC is a product function or process step.

Typically development teams think of nonmeasurable SQCs as either "present" or "absent." When determining an impact in this context, they try to judge whether customer satisfaction performance will be high if the SQC is present, and low if the SQC is absent. If the range in customer satisfaction performance level is predicted to be large, the team will assign ◎ as the impact.

In fact, it is generally unrealistic and counterproductive to think of product features or service elements in this binary fashion. Let's take an example from a software development application. Consider the product function "File OPEN command" and its linkage to the customer need:

CAN MIX MATERIAL FROM MANY DOCUMENTS

A software developer would find the question

> What is the linkage between **CAN MIX MATERIAL FROM MANY DOCUMENTS** and "File OPEN command"?

meaningless, because the software must include the "File OPEN command" in order to be functional at all. In other words, at first glance, since there must be a "File OPEN command" in any case, there is no need to ask the question. Lurking behind the developer's response is the assumption that the only choice is "provide an OPEN command" and "don't provide an OPEN command."

In fact, there is a continuum of choices. For example, the OPEN command function could be designed to allow the user to open many documents at once, thereby strengthening the linkage to customer satisfaction. Or it could be designed to open only one document, thus forcing the user to invoke the command several times in order to access several documents and thereby lowering customer satisfaction performance for this need.

Another variation in the OPEN command design might be to provide the user with a convenient list of files that have been associated with any currently opened or recently opened files.

Many other variations on the possible capabilities of the OPEN command could be imagined by the creative software engineer. These creative options do not occur on a simple linear continuum. However, the development team would do well to create a model in their minds of a simple continuum of design possibilities ranging from "stripped down OPEN command" to "deluxe OPEN command." This model provides the team with a realistic range of possible technical responses for the SQC. If several high linkages show up for a product function, then a "deluxe" version of that function may be called for.

8.2   Impact Values

In Diagram 8-2 we have seen the most common numerical values that QFD teams assign to the linkage symbols. Some early QFD applications in the U.S. used 5, 3, 1, 0 for the impacts from "strong" to "none." Over time, QFD facilitators felt the need to create a stronger contrast between "strong" and the other impacts, so that strong impacts would have more influence on the final prioritizations. The value 9, which Don Clausing had seen in Japanese applications in the early eighties, was rapidly adopted for this purpose. "9" is a comfortable multiple of "3" and serves the purpose of making the "strong" impacts dominate the matrix.

Some facilitators favor "7," because it is a compromise between "5" and "9." Others favor "10" because it serves much the same purpose as "9" but provides for easier manual calculations than any of the other choices.

The greater the ratio between the values assigned to "strong" and "moderate," the less likely it is that an SQC with only "moderates" assigned to it will have a technical importance greater than an SQC with at least one "high." Most QFD practitioners in the U.S. feel that is the way things should be.

There is no scientific basis for any of the choices. The impact values simply provide a way for the development team to express its judgment on the relative impacts of SQCs on customer needs. The SQCs can then be differentiated in terms of their overall contribution to customer satisfaction performance.

8.3   Priorities of Substitute Quality Characteristics

Once the development team has determined all the impacts or linkages, some simple arithmetic provides one of the key results of QFD: the relative contributions of the Substitute Quality Characteristics to overall customer satisfaction.

These represent the priorities of the SQCs, and are placed near the bottom of the house as shown in Diagram 8-5. Diagram 8-6 illustrates how it works.

The impact of Technical Response X to Need A is "high." We multiply the numerical value for "high" (9) by the Normalized Raw Weight for Need A (.43). The result of 3.9 has been written above the diagonal of the cell. This value is called the *Relationship* of Technical Response X to Customer Satisfaction Performance on Need A. After computing all relationships, we add all the relationships for a Substitute Quality Characteristic and put the result in a Totals row at the bottom of the matrix. These Totals are called the *contributions* of the SQCs to overall customer satisfaction. The larger the contribution, the more influence the SQC has on Customer Satisfaction Performance, and therefore the more important it is for the product or service to do well in the implementation of that SQC.

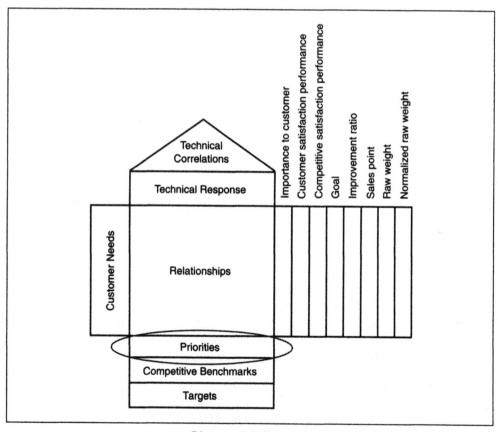

Diagram 8-5. Priorities

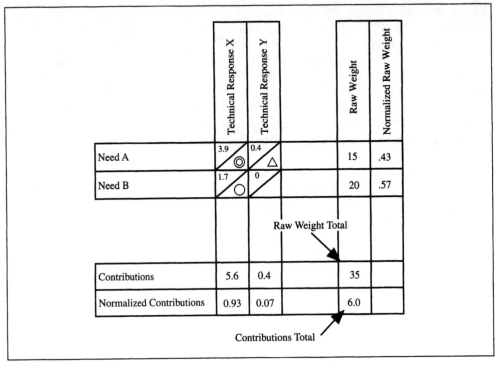

Diagram 8-6. Contribution Calculations

If the SQCs and their contributions are to be transferred to the left side of another matrix (as in Diagram 3-12), it is useful to convert the contributions to normalized contributions, using a normalization method analogous to the one described in Section 6.7, and to use the normalized contributions in the next matrix.

8.4   Negative Impacts

It happens occasionally that a Substitute Quality Characteristic is found to have a *negative* impact on customer satisfaction performance for a certain attribute. In other words, as the SQC is conceptually rotated in the direction of goodness, customer satisfaction performance for a certain attribute is judged by the development team to go down.

Such negative relationships can happen when an SQC has been selected for its positive impact on one or more attributes, but the attribute with which it is negatively linked has not yet been considered. Some examples are the following:

In computers, faster internal clock speed may have a positive impact on the customer's need to get work done faster, but it may also have a negative impact on the customer's need for system reliability. Faster clock speed implies higher internal operating temperatures, which generally cause parts to deteriorate faster.

In service, a wider range of services may have a positive impact on the customer's need for getting all transactions processed from a single source, but it may have a negative impact on the customer's need for a simple method to use the entire range of services (since more choice implies more difficulty in making a selection).

In automobiles, thicker steel sheets on the doors may have a positive impact on the customer's need for safety in case of collision, but may have a negative impact on the customer's need for good fuel efficiency (since the weight of the car would go up with thicker steel sheets).

These negative impacts complicate QFD computations. They also require some care on the part of the development team when analyzing the QFD results. Here is a good way to handle negative impacts:

1. Define special symbols to represent the negative impacts—for example, $-\textcircled{\small\bullet}$, $-\text{O}$, and $-\triangle$.
2. Define corresponding numerical values for the negative impacts (-9, -3, and -1 for strong negative impact, moderate negative impact, and possible negative impact are most common). Multiply the appropriate impact by the Raw Weight to calculate the relationship (as described in Section 8.3). Along with the positive relationships, some of the resulting relationships will now be negative.
3. For each Relationships Section column containing negative impacts, compute *two* sums: the algebraic (signed) sum of the relationships, and the sum of the absolute values of the relationships.
4. If the difference between the algebraic total and the absolute value total is small (as for Technical Response X in Diagram 8-7), then the effect of the negative impact is small and can probably be disregarded.
5. If the difference between the algebraic total and the absolute value total is large (as for Technical Response Y in Diagram 8-7), then the effect of the negative impact cannot be ignored. The team is being confronted with a "breakthrough opportunity." That is, the team is being challenged to define one or more Technical Responses that provide positive impacts only to replace the one that contains negative impacts. One way to try to solve this problem is to define Technical Responses that each have a narrower range of impact.

An alternative method for dealing with negative impacts, favored by some QFD facilitators, is

1. Express all impacts on customer satisfaction performance as positive.
2. Study the negative impacts as they are reflected in the Technical Relationships (the roof) section of the HOQ (Chapter 9).

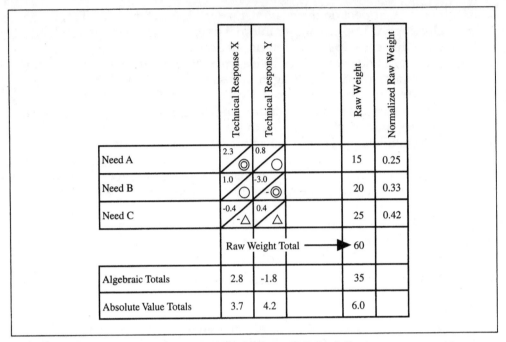

	Technical Response X	Technical Response Y		Raw Weight	Normalized Raw Weight
Need A	2.3 ◎	0.8 ○		15	0.25
Need B	1.0 ○	-3.0 -◎		20	0.33
Need C	-0.4 -△	0.4 △		25	0.42
	Raw Weight Total ⟶			60	
Algebraic Totals	2.8	-1.8		35	
Absolute Value Totals	3.7	4.2		6.0	

Diagram 8-7. Negative Impact Calculations

Regardless of the approach, it's better to get rid of negative impacts than to find ways of handling them in QFD. If negative impacts show up in the QFD process, try to find new SQCs that have a positive impact across all customer needs.

8.5   Many-to-Many Relationships

As a final point, we should note the process we use in QFD to finally arrive at the contributions of the SQCs. We normally start with Customer Needs. From each need we generate one or a few SQCs (see Diagram 7-4). We might expect that when we determine the impacts, we will see high impact for SQCs that were generated to relate strongly to the need from which they originated, and low impacts elsewhere—basically a one-to-one relationship between SQCs and Customer Needs, as in Diagram 8-8.

In practice, the SQCs tend to take on a life of their own once they are placed in the matrix. Their impact on *all* customer needs must be evaluated, and many of the surprises of QFD spring from these evaluations. The SQCs invariably are found to relate to many customer needs, sometimes even more strongly to customer needs that did not suggest the SQCs in the first place (see Diagram 8-9).

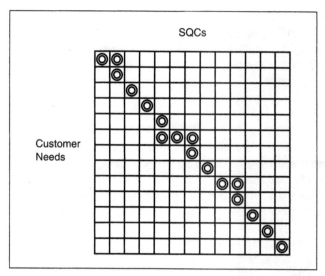

Diagram 8-8. Pattern of Impacts after SQCs Have Been Deployed from Customer Needs (One-to-One)

Unexpected strong and weak relationships come up. SQCs thus emerge as important because of relationships with customer satisfaction that no one had previously understood.

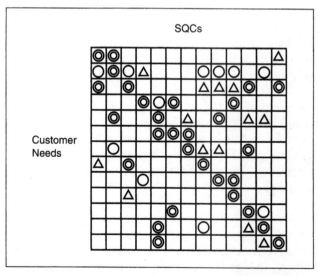

Diagram 8-9. Pattern of Impacts after All Relationships Have Been Evaluated (Many-to-Many)

It is the matrix structure of QFD that helps the team systematically evaluate all possible relationships, even those that "could not possibly matter." Thus, we see how the QFD process helps a team to ask questions it might not otherwise ask, and therefore arrive at answers it might otherwise overlook.

Summary

We have now seen how the elements of the Technical Response (the Substitute Quality Characteristics) can be prioritized. It is done by estimating the impact of each SQC on the customer satisfaction performance of each Customer Need. The Relationships section of the HOQ provides the mechanism for displaying all the relationships, and for computing the priorities of each SQC.

The Relationships section lies at the heart of QFD. It provides the development team with a way of mapping the relevance of proposed technical responses to customer needs.

The development team makes a determination, for each SQC/Customer Need pair, of the impact that SQC has on customer satisfaction performance on that Customer Need. The impacts are usually represented by symbols. The numbers related to these symbols are a measure of the amount of impact.

The product of the impact with the importance of the Customer Need (the Need's Raw Weight) is called the Relationship. The Relationship is a measure of the importance of the impact to overall customer satisfaction.

The sum of Relationships for a single SQC is called the Contribution of that SQC to overall satisfaction. The relative values of Contributions for all SQCs provides a rank ordering of SQCs that can guide trade-offs and resource allocation for the remainder of the development process.

Judging impacts is never easy, but it is especially difficult when the SQC has not been expressed as a continuous variable. Nevertheless, for nonnumerical SQCs, there are ways of thinking of them as ranging from "basic" to "deluxe," which help to define a continuum of effort or attention. This continuum makes it easier for the team to judge impact, and also to develop creative solutions to meeting customer needs.

Now that we understand how the SQCs relate to customer needs, we have to ask how the SQCs relate to each other. Is it possible that performing well on one SQC might cause us to perform poorly on another? Or perhaps performing well on one SQC makes it easier to perform well on another? The Technical Correlations chapter shows us how to do this analysis.

Discussion Questions

Take a subset of your known customer needs. Create an abbreviated House of Quality matrix and place these needs on the left. Put a few of your Substitute Quality Characteristics on top of the matrix. Fill in the impacts according to your best judgment. Any surprises?

Show your matrix to a colleague. Do you both agree on the way you have set up the matrix, and on its contents? If you don't agree, what can you learn from each other?

CHAPTER 9

Technical Correlations

Introduction

We have mentioned earlier that QFD is a key to concurrent engineering because it facilitates team members communicating with each other. We are about to see in this chapter an explicit mechanism for mapping out exactly what kind of communication must occur on the development project. The Technical Correlations section (Diagram 9-1) provides the mechanism. It will show us for which technical areas close communication and collaboration are important, and for which it is not. It will also show us where design bottlenecks may occur, and therefore where design breakthroughs are needed.

The Technical Correlations section is probably the most underexploited part of the House of Quality. Few QFD applications use it, yet its potential benefits are great. Perhaps after reading this chapter, you will be encouraged to use it yourself for the competitive advantage it offers.

9.1   Meaning of Technical Correlations

The Correlations section of the House of Quality is sometimes called the Technical Correlations section. More often, it is referred to as the "roof" of the House of Quality. It maps interrelationships and interdependencies between

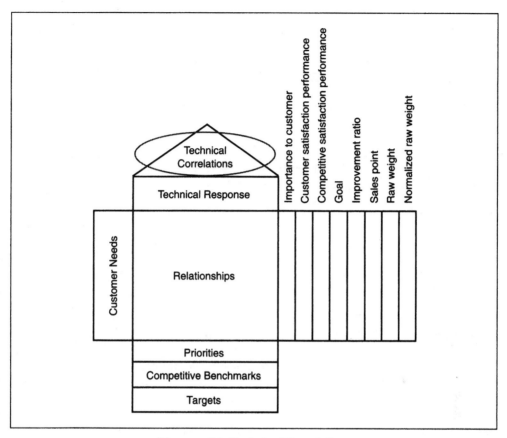

Diagram 9-1. Technical Correlations

Substitute Quality Characteristics. This section of the House of Quality is probably the least used in today's practice of QFD. However, as this chapter indicates, the analysis of the roof can lead to important insights in the development process.

The Correlations section consists of that half of a matrix that lies above the matrix's diagonal. The SQCs are arrayed along the top and side as shown in Diagram 9-2. The matrix is then rotated 45 degrees, and since the SQCs are already available along the top of the HOQ, they double as the labels for both the rows and columns of the roof, making the row and column labels unnecessary (see Diagram 9-5).

Very often, especially after a technical concept has been decided upon and is somewhat understood, the developers will be able to see that as SQC_x is

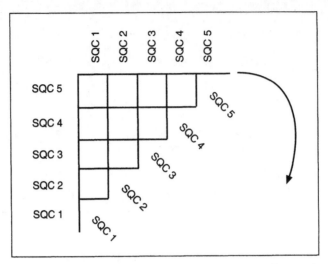

Diagram 9-2. Matrix above the Diagonal, to Be Rotated

moved in the direction of goodness, SQC_Y will be influenced, either in its direction of goodness or in the opposite direction. The degree and direction of influence can have a serious impact on the development effort. Indications of negative impact of one SQC upon another represent bottlenecks in the design. They call for special planning or breakthrough attempts.

For example, in a case where a producer of integrated circuits was developing an "ASIC" (Application-Specific Integrated Circuit) for a customer, analysis of the Technical Correlations disclosed that certain customer requirements were technically incompatible. Had this incompatibility not been discovered during the product planning phase, the ASIC producer believes they would have wasted several million dollars in preliminary development work before discovering the need to redesign.

Other examples of conflicting technical correlations are the following:

- For an automobile, increased BTU rating of an automobile air conditioner (SQC_X, more is better) may have a negative impact on automobile weight (SQC_Y, less is better). Notice how the SQCs are somewhat solution-dependent, because it is assumed that an air conditioner will be used to cool the auto interior, and that higher BTU ratings imply heavier air-conditioning units.
- In computer software, an increased number of print command options (SQC_X, more is better) may have a negative impact on the number of key strokes and mouse clicks required to invoke a print command (SQC_Y, less is better).

- For service, reducing the number of minutes a customer speaks to a service associate (SQC_X, less is better) may have a negative impact on the number of calls a customer must make to get a problem solved (SQC_Y, less is better).

In QFD we usually identify five degrees of technical impact, as in Diagram 9-3.

These symbols carry no directional connotation. Some QFD practitioners believe that the technical correlations should be treated as bidirectional in impact. Then, if either SQC varies, the other will vary according to the type of correlations described between them.

The author feels it's more constructive to indicate a direction of impact, since a developer can often make a strong argument for impact of SQC_X upon SQC_Y, but no impact of SQC_Y upon SQC_X.

For example, for the automobile, increased BTU rating of an air conditioner could raise automobile weight, but increased automobile weight does not necessarily affect the air conditioner's BTU rating. The counter-argument is that increased automobile weight *does* affect the air conditioner's BTU rating, because the air conditioner will have to be more powerful to achieve the same cooling in a heavier car as compared to a lighter car.

If the QFD team wishes to record direction of impact, the symbols in Diagram 9-3 can be coupled with an arrow indicating the direction, as in Diagram 9-4. Bidirectional impact can be denoted by a two-headed arrow (⟷).

Once the Technical Correlations matrix has been rotated, the redundant row and columns removed, and correlations filled in, it might look like Diagram 9-5. The interpretation of the topmost cell, for example, is then

"Moving SQC 1 in the direction goodness has a moderate negative impact on SQC 5's direction of goodness."

✔✔	Strong positive impact
✔	Moderate positive impact
<blank>	No impact
✕	Moderate negative impact
✕✕	Strong negative impact

Diagram 9-3. Degrees of Technical Impact

$\overset{\longrightarrow}{v\!v}$	Strong positive impact, left to right
$\overset{\longleftarrow}{v}$	Moderate positive impact, right to left
<blank>	No impact
$\overset{\longleftarrow}{x}$	Moderate negative impact, right to left
$\overset{\longrightarrow}{xx}$	Strong negative impact, left to right

Diagram 9-4. Degrees of Technical Impact with Direction of Impact

9.2   Responsibility and Communication

One of the most important benefits of the Technical Correlations is to indicate which teams or individuals must communicate with each other during the development process. If a team with prime responsibility for meeting the target value of SQC 1 runs into difficulties, or changes plans, Diagram 9-5 tells this team that the team responsible for SQC 4 will be seriously affected, and the team responsible for SQC 5 will be somewhat affected, by their change of plans.

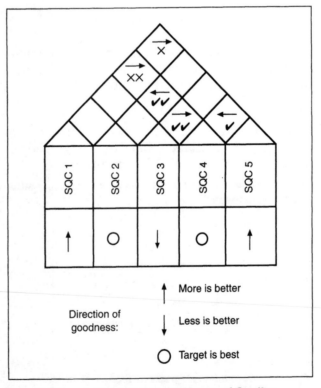

Diagram 9-5. Roof of the House of Quality

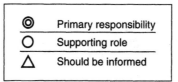

Diagram 9-6. Responsibility Symbols

One method for making this information more explicit is to construct a *Responsibility Matrix*. This matrix would display the SQCs along the left, and the possible responsible teams along the top. A cell in the matrix would indicate the relationship of the team to the SQC. The possible relationships are shown in Diagram 9-6. A good management practice is to assign responsibility for an objective, such as meeting the target value for an SQC, to a single individual or a single organization. Therefore, the responsibility matrix would have a single ◎ in each row.

A responsibility table for the roof shown in Diagram 9-5 might look like Diagram 9-7.

In Diagram 9-7, Organization A has primary responsibility for SQC 1. Because changes in SQC 1 strongly affect SQC 4, the organization responsible for SQC 4 (Organization E) must be informed of progress on SQC 1. Organizations C and

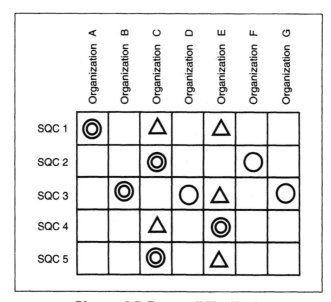

Diagram 9-7. Responsibility Matrix

F have primary and support responsibility, respectively, for SQC 2. SQC 2 has
no impact on any other SQC, however, so communication about SQC 2 progress
does not have to be communicated to any other organizations.

9.3   Correlations Network

An alternative but equivalent representation of the correlations in the roof is
the Relationship Digraph, or Relationship Network diagram, shown in Diagram 9-8. In this diagram, the SQCs are represented by the circles, and the
SQC affected is shown by the arrows connecting the circles. The degree and
direction of influence is shown by the ✔ and X indications written alongside
the arrows. A more elaborate formulation of this type of network, called the
Interpretive Structural Model, has been developed by Professor John N. Warfield,[1] but is beyond the scope of this book.

Notice that SQC 1 in Diagram 9-8 has arrows emanating from it, and none
entering into it. This is an indication that SQC 1 is a *driver* in the sense that it
influences other SQCs but is not in turn influenced by any SQCs. On the other
hand, SQC 2 has only incoming arrows and no outgoing ones. It is called an
indicator. The status of SQC 1 as a driver indicates that efforts to move SQC 1
in the direction of goodness will affect other SQCs as a byproduct (in this
case, negatively). It does not seem worthwhile to invest resources in moving
SQC 2 directly, since it is strongly affected by SQC 4 directly, and by SQC 1, 5,
and 3 indirectly.

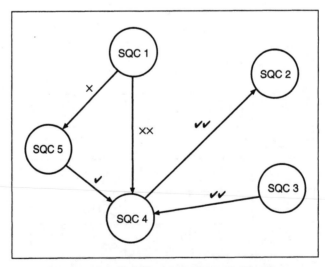

Diagram 9-8. Relationship Network Diagram

Summary

The "roof" of the House of Quality shows the impact of work on one SQC on the status of other SQCs. The roof can show the existence and nature of design bottlenecks. Where bottlenecks or opposing SQCs exist, the team must plan a concentrated activity to accomplish both.

The correlations in the roof spell out which organizations or individuals must communicate with which others, to allow for smooth progress in meeting target values of SQCs during development. To ignore these interrelationships is to invite chaos into the development process, because changes in plans in one area may go unnoticed in another affected area until too late in the game. Without the analysis inherent in the roof, some interrelationships will almost certainly be ignored.

Communication about, and responsibilities for, SQCs can be summarized in a Responsibility Matrix. This matrix provides a clear graphical map of ownership for the SQCs.

An alternative graphical representation of the correlations is the Relationship Digraph. This diagram helps to identify the drivers and indicators among the SQCs.

We have now prioritized the SQCs and evaluated their interrelationships. It's time to set technical targets. But technical targets cannot be set in a vacuum. We must see how well the competition is doing and set targets that will ensure that we are competitive. This implies benchmarking the competition. We'll do this in the next chapter.

Discussion Questions

Given the key Substitute Quality Characteristics in your development work, which are the drivers? Which are the indicators? How did you determine your answer?

Who needs to be informed of design changes? What analysis have you done that shows who should be informed?

---

[1] John N. Warfield, "A Science of Generic Design: Managing Complexity Through System Design, Second Edition," Iowa State University Press, 1994.

CHAPTER 10

Technical Benchmarks

Introduction

This chapter introduces the concept of competitive technical benchmarking (Diagram 10-1).

No organization would invest in the development of a product or service without knowing enough about the competition to be sure that their design is competitive. But if we have defined thirty or forty SQCs, should we benchmark all of them? Because they have been prioritized using the Relationships section, as we have seen in Chapter 8, we know which SQCs are most important, and we therefore have a strategy for focusing on and benchmarking only the most important of them.

In this chapter we'll look at the process of benchmarking two types of SQCs: those that have been formulated as performance measures, and those that have been formulated as product functions.

Once the SQCs have been prioritized, the next step is to set targets for them. The most important ones—those with the highest contribution value—require the most care. The QFD analysis indicates that performance of the most important SQCs will strongly influence customer satisfaction, so the

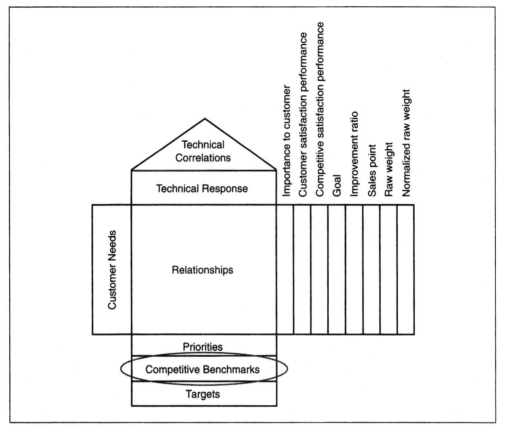

Diagram 10-1. Competitive Benchmarks

development team will want to set targets for those SQCs aggressively. Subsequent management of the development process will then be oriented to assuring success in meeting those aggressive targets.

A critical question is: how should the targets be set? How aggressive do they need to be? To a great extent, development teams can be guided at this point by the competition's performance as well as their own performance. If the team knows how well they and the competition are currently performing on the most important SQCs, they can make crucial strategic decisions whether to match the competition's performance, exceed the competition's performance, or even concede technical superiority to the competition. We'll discuss target setting in more detail in Chapter 11. But before setting targets, the development team would do well to study their competition by benchmarking.

The QFD process provides the basis for strategic competitive benchmarking. The highest ranking SQCs are the ones that determine the success of the product or service; thus, they are the ones to be examined in competitive offerings.

In the process of examining them, the language of the Substitute Quality Characteristics and the definition of direction of goodness become important determiners of the competitive benchmarking work.

What do we mean by competitive benchmarking? In general, competitive benchmarking is the process of examining the competition's product or service according to specified standards, and comparing it to one's own product or service, with the objective of deciding how to improve one's own product or service. In QFD terms, the standards are defined as the most important SQCs, as the team defined them earlier in the QFD process.

10.1   Benchmarking Performance Measures

If the SQCs were defined as performance measures, the benchmarking process becomes one of measuring the competition's performance and one's own performance in terms of these measures. To the extent that the performance measures were defined independently of the design of the product or service, the benchmarking process provides ideal "apples-to-apples" comparative data between the competition's and the development team's product or service. The results of measuring the two products or services can be laid down side by side (in the HOQ, one above the other) and evaluated at a glance, as in Diagram 10-2.

Some teams prefer graphical benchmark comparisons, as in Diagram 10-3. This is similar to the graphical representation of customer satisfaction performance described in Section 6.3 (Diagram 6-14).

10.2   Benchmarking Functionality

If the SQCs were defined in a more solution-specific manner, with product or service functions explicitly defined, the comparisons must be much more subjective. This is because the competition's design of a particular SQC is likely to be different, at least in some respects, from the development team's. One way to deal with these differences in functionality is to decompose the high-ranking SQCs into "sub-SQCs" and compare these "sub-SQCs" to the competition's.

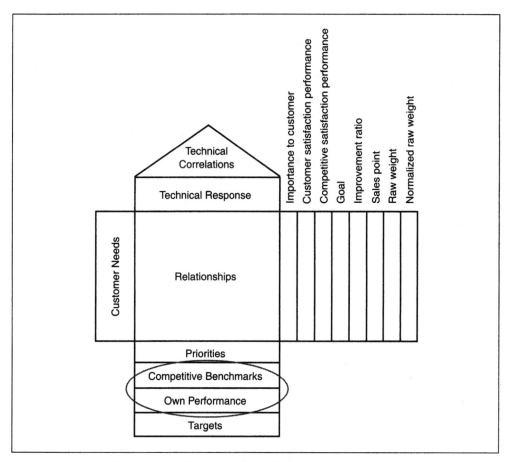

**Diagram 10-2. Comparison of Competitive Benchmarks
and Company's Own Performance**

If the process of developing the SQCs in the first place was by the Function Tree method or the Affinity Diagram method, then the sub-SQC's come from the lower levels of the Function Tree of the Affinity Diagram. In Diagram 7-5, we see the explosion of an SQC into a long list of subfunctions. These subfunctions could be compared to the competition's subfunctions. Some subfunctions may correspond very closely to the competition's; others will not.

The number or percentage of subfunctions that correspond provides valuable numerical information. A listing of the functions that do not correspond provides data for understanding where the competition provides more functionality, or how the competition's design solves the same problem differently from the development team's design.

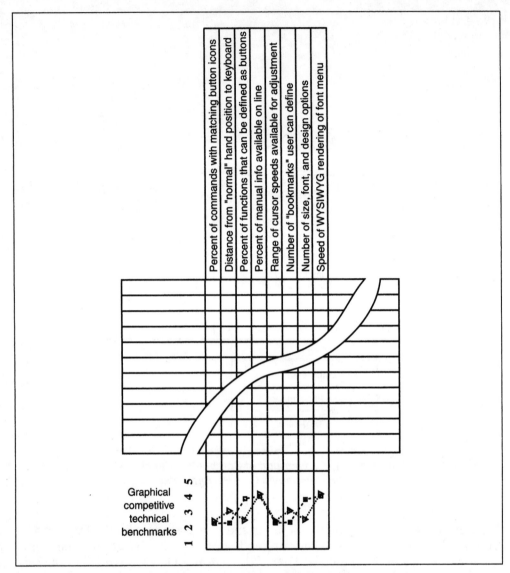

Diagram 10-3. Competitive Benchmarks in Graphical Display

Summary

In Chapter 10 we have seen how QFD provides a method for benchmarking strategically: we use the priorities of the SQCs, determined by completing the Relationships section, to guide us in selecting which SQCs to benchmark. The language of the comparison should be dictated by the language used to define the SQCs.

We can benchmark SQCs described as performance measures, and, with somewhat less precision, we can benchmark SQCs expressed as product or service functions.

Now that we have measured the competition, we are in a position to set target values for our key SQCs. That's the topic of the next chapter.

Discussion Questions

How is competitive benchmarking done in your organization? How do you decide what aspects of the competition's offerings to measure or examine? What is the linkage between what you benchmark and what your customers' needs are? Compare your method of deciding what to benchmark to the way QFD helps you decide.

What language is used for making comparisons? Is it numeric? If not, how does it compare to the language used for writing requirements, specifications, and design documents? How effective is it in providing side-by-side comparisons between your products or services and the competition's?

How much money should be spent on competitive benchmarking? How much does your organization spend?

CHAPTER 11

Targets

Introduction

This chapter covers the process of setting targets for the key SQCs (Diagram 11-1). It takes up the QFD process after the development team has determined the most important SQCs and has benchmarked the competition.

We will look at setting numerical targets for SQCs that have been expressed as performance measures, and at setting function or feature targets for SQCs that have been expressed as features. In the case of numerical targets, we'll look at the possibility of using some simple algebra as a guide. For features, we'll make use of the idea of expressing features in terms of their "subfeatures," as we discussed in Section 10.2.

Setting targets is of course a matter of greatest interest to product and service developers. Obviously, setting SQC targets will drive all subsequent development activity. Development teams set targets for themselves whether or not they use QFD to plan their project. Without a process such as QFD, the targets tend to be a hodgepodge of customer-oriented and technical goals, with very little linkage or relationship among them. Nor is the prioritization of these targets based on a line of reasoning that all others can follow, let alone agree with.

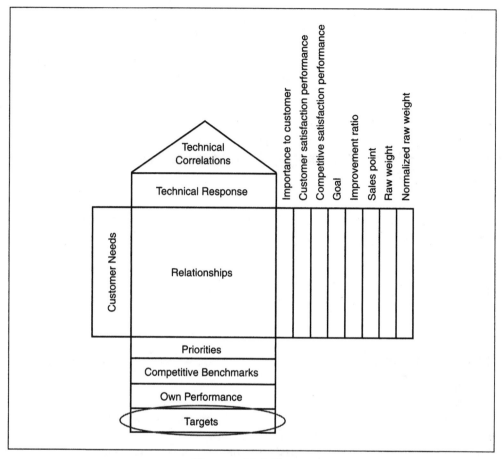

Diagram 11-1. Targets

With QFD, the targets have a context: they are related to customer needs, to the competition's performance, and to the organization's current performance. The rank ordering of the targets is based on the systematic analysis done in the Relationships section (and all the prior QFD analysis). The rank-ordering process is traceable, because all the decisions affecting the rank ordering are recorded in the QFD matrix.

The QFD process itself provides no cookbook approach for setting targets for SQCs. The most vital information not explicitly visible in the House of Quality is the business know-how and technical expertise of the development team. However, the House of Quality provides much of the strategic information needed, laid out in a compact form.

In Section 8.1.1 we referred to the possibility that some of the linkages in the Relationships section might not be linear, because the associated SQCs may be Dissatisfiers or Delighters. The target-setting stage is a good time to deal with these Kano classifications. It is useful at this point, if the team has not already done so, to classify the SQCs according to their category in the Kano model.

For those SQCs classified as potential Delighters, the team must decide how aggressive it can afford to be in target setting. There is relatively little downside risk in setting a conservative goal: customers will not notice the absence of a Delighter. However, the potential gain of setting a goal that beats the competition is high.

For those SQCs classified as Dissatisfiers, the team cannot afford not to be aggressive. Customers are expecting perfection in these areas, and anything less than perfection will result in customer dissatisfaction.

For those SQCs classified as Satisfiers, the team can expect that the better they perform on the SQC, the greater the customer satisfaction performance will be for the linked customer needs. The rest of this chapter deals with strategies that can help in setting targets for Satisfiers.

11.1   Numerical Targets

11.1.1   Comparison with Competition

One approach to setting targets is similar to the process of setting customer satisfaction performance goals in the Planning Matrix (Section 6.4). The primary inputs to goal setting for customer satisfaction performance in the planning matrix are

- Importance of customer attribute to customer
- Our current satisfaction performance rating
- Competition's satisfaction performance rating

Similarly, the primary inputs to target value setting of Substitute Quality Characteristics are

- Rank order of Substitute Quality Characteristics (Priorities)
- Competition's technical performance (Competitive Benchmarks)
- The development team's product's technical performance (Own Performance)

The line of reasoning for setting targets is also similar to that used in setting goals in the planning matrix. Starting with the highest ranking SQC, determine the strength of the development team's position relative to that of the competition. Based on the team's knowledge of the difficulty of performing well on the SQC, the team can decide whether to aim to do better than the competition, to match the competition, or to concede technical leadership to the competition. As a general rule, the goal should be for technical performance that exceeds the best in the world for those SQCs that matter the most to overall customer satisfaction.

11.1.2 Mathematical Modeling

While QFD is certainly not a precise mathematical model of the relationship between technical performance and customer satisfaction performance, a little bit of simple mathematics could serve as a guide to the development team in setting targets.

In the case of a Substitute Quality Characteristic for which Less Is Better, we may imagine that the relationship

Customer Satisfaction Performance$_A$ = f(Technical Performance Measure$_x$)

is approximated by a linear function of the form $y = m \cdot x + b$. The slope m of the line is defined by two known points.

The coordinates of the first point are ($p_{world\ class}$, $s_{world\ class}$)—that is, the point that corresponds to the best product in the world. Satisfaction performance s is highest, and technical performance p is lowest (best). In Diagram 11-2, customer satisfaction performance for this point was measured at 4 (with 5 as the best possible score), and technical performance was determined by competitive benchmarking to be 80.

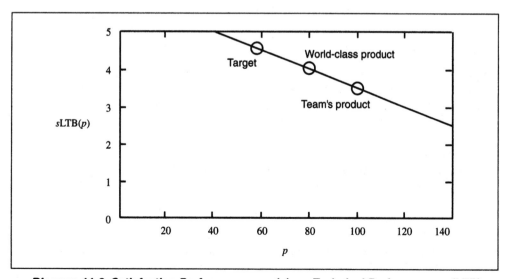

Diagram 11-2. Satisfaction Performance s$_{LTB}$(p) vs. Technical Performance p (LTB)

The coordinates of the second point are (s_0, p_0)—that is, the point that corresponds to the development team's current product. Satisfaction performance was determined by market research to be modest (3.5) and technical performance (based on laboratory measurements) was higher than (not as good as) that of the world-class product (100).

Visual inspection of the line passing through these two points in Diagram 11-2 suggests that a good value of p to aim for might be 60 or less, which could result in a customer satisfaction performance level of about 4.5.

The general equation for the line is

$$s_{LTB}(p) = \frac{s_{\text{world class}} - s_0}{p_{\text{world class}} - p_0} \bullet p + \left[s_0 - \left(\frac{s_{\text{world class}} - s_0}{p_{\text{world class}} - p_0} \right) \bullet p_0 \right]$$

where:

$s_{LTB}(p)$	denotes customer satisfaction performance on a customer need as a function of a Substitute Quality Characteristic p of the type Less the Better
$s_{\text{world class}}$	denotes customer satisfaction performance with the best product in the market
s_0	denotes customer satisfaction performance with the development team's product
$p_{\text{world class}}$	denotes technical performance of a Substitute Quality Characteristic with the best product in the market
p_0	denotes technical performance of a Substitute Quality Characteristic with the development team's product
p	denotes technical performance of a Substitute Quality Characteristic

A similar but slightly more complex relationship can be modeled in the case of Target Best (TB). Here we might consider that a parabola best describes the relationship between technical performance and customer satisfaction performance, as in Diagram 11-3, where the target value for p is 4 (the meaning of the other variables is as before).

The coordinates of the first point are $(p_{\text{world class}}, s_{\text{world class}})$—that is, the point that corresponds to the ideal target value. Since the direction of goodness is Target Is Best, it is not possible to perform better than the target (4). In this example, we have assumed that achieving the target (4) results in the best possible customer satisfaction performance (5).

The coordinates of the second point are (s_0, p_0)—that is, the point that corresponds to the development team's current product. As in the previous example,

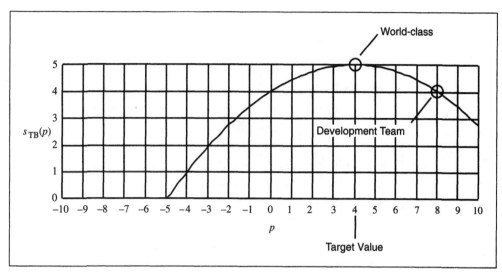

**Diagram 11-3. Customer Satisfaction Performance s(p)
vs. Technical Performance (p), Target Is Best**

customer satisfaction performance was determined by market research, and the value of the SQC (4) was determined by laboratory measurement.

The general equation for the parabola in Diagram 11-3 is

$$s_{TB}(p) = \frac{\left| s_0 - s_{\text{world class}} \right|}{\left(-p_0^2 + 2 \cdot p_0 \cdot \text{target value} - \text{target value}^2 \right)} \cdot (p - \text{target value})^2 + s_{\text{world class}}$$

This mathematical model provides some insight for setting the target for a Substitute Quality Characteristic, but care must be exercised in its use. For one thing, there is no guarantee that the relationship between any SQC and the corresponding customer attribute satisfaction performance is precisely linear or precisely quadratic. More importantly, customer satisfaction performance for any attribute is usually not the function of a *single* SQC, but of several SQCs.

Consider the Relationship section matrix in Diagram 11-4. The relationship between SQC X and Attribute D has been modeled as one-to-one. The team could consider that satisfaction performance on Attribute D is solely related to SQC X. However, what can the team assume about the relationship between SQC U and Attribute A? Two other SQCs (V and Z) besides SQC U contribute to satisfaction performance on Attribute A. The relationship, if it can be modeled at all, is much more complex than a simple one-to-one function. Even if

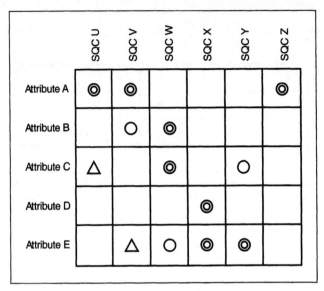

Diagram 11-4. Relationship Section

the development team assigned one-third of the influence on customer satisfaction performance to each SQC, other complications could muddy the waters. For example, the direction of goodness might not be the same for each SQC.

Since there are no reliable multivariate models for setting target values in QFD, a reasonable approach could be the following:

1. Treat each SQC *as if* it were the only SQC contributing to customer satisfaction performance of an attribute.
2. Use the simple models described here to create a first estimate of an appropriate target value.
3. Repeat this analysis for all the customer satisfaction attributes that this SQC is linked to. This creates multiple target values for the same SQC. In Diagram 11-4, for instance, we would estimate three target values for SQC W: one each for its relationships to Attributes B, C, and E.
4. Choose the most aggressive of these target values.

11.2   Nonnumerical Targets

Setting targets for Substitute Quality Characteristics defined as features or processes is obviously more difficult than dealing with numbers. A number is one-dimensional, but features and processes are multidimensional and multi-faceted.

There are two helpful ways of thinking about targets for nonnumerical SQCs: the continuum model and the subfeature model.

In the continuum model (Diagram 11-5), we may imagine the SQC to be on a continuum as described in Section 8.1. This continuum could have as its endpoints "stripped down" and "deluxe." The development team could judge where on the continuum their current offering lies, and where the best in the world lies. To clarify these judgments to themselves and to others, they would do well to make their subjective judgments as objective as possible by documenting:

- The differences between "Best in world" and "Deluxe"
- The differences between "Development team's" and "Best in world"
- The differences between "Development team's" and "Target"

By using the subfeature model, the development team can explode each feature to be targeted into its component subfeatures, as described in Chapter 7. Each subfeature could be evaluated according to the continuum model, or could be exploded into lower-level subfeatures. Targets could be set by continuum, where "Best in world" and "Development team's" subfeatures line up, and by identifying subfeatures to be added, where the features don't line up. In Diagram 11-6 we can see at a glance how the subfeatures line up, and how those that are present score via the continuum model. The arguments for setting targets are then easier to justify and explain. The column of continuum targets provides a profile of the target that is at least comparable in detail to the profiles of the "Development team's" and the "Best in world" profiles.

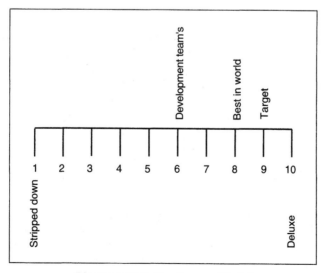

Diagram 11-5. Continuum Model

	Development team's	Best in world	Target
Subfeature a	7	7	8
Subfeature b		5	4
Subfeature c	6	8	8
Subfeature d	3		3
Subfeature f		8	
Subfeature g	6		6

Diagram 11-6. Target Setting by Subfeature and Continuum Analysis

In Diagram 11-6, the development team's existing product, and the Best in World product, both deliver Subfeature a at about the same level on the Continuum model. The team has decided to set a target for this subfeature to outperform the Best in World. In the case of Subfeature d, the development team's level of delivery on Subfeature d is not very high (3), but the Best in World lacks this Subfeature. Assuming the Best in World will not provide Subfeature d within the product planning time frame, the team has decided to continue delivering the subfeature, but not to improve or expand it. Hence, the Target has been set to the same value at which the subfeature is already rated. In the case of Subfeature f, the Best in World already provides that subfeature at a high level. The development team has decided to concede this subfeature to the competition, and use its resources to compete in other areas.

Summary

In this chapter we have seen how to set targets for performance measures and how to set targets for functions or features. Target setting is easier and more objective when Substitute Quality Characteristics are expressed as performance measures. Nevertheless, even with SQCs that are expressed nonnumerically as features or processes, it is possible to set targets in terms of "degree of deluxeness" of implementation, and in terms of the existence of subfeatures.

In any case, setting targets for the most important SQCs provides the development team with a systematic method for deciding how to compete against the best in the world.

This completes our description of the elements of QFD. We have explored all the sections of the House of Quality in detail. We have seen how QFD encourages us to listen to the customer and to represent customer needs in a hierarchical format—a particularly useful representation in which we can zoom in or zoom out, and in which we can understand the relative importance of each customer need, as well as see how well we are performing in meeting each need.

We have seen in the Planning Matrix how to use the customer needs and competitive market research to make strategic top-level decisions about what to emphasize in the development project. The primary result of this strategic planning is the Raw Weights that represent the revised relative importances of the customer needs, based on a combination of customer perceptions and the company's business objectives.

We have explored various ways of generating a statement of Technical Response—the Substitute Quality Characteristics (SQCs) that together represent the product or process requirements. The primary approach, but not the only possible approach, has been to "deploy" the customer needs into a set of solution-independent Performance Measures.

We have explored QFD's unique method for establishing the linkage between the Technical Response and the Customers' Needs—the development of impacts in the Relationship section.

Finally, we have examined the QFD approach to prioritizing the elements of the Technical Response by multiplying the impacts by the Normalized Raw Weights of the customer needs and adding them for each SQC. We have also seen how these priorities can be used to shape a benchmarking effort, and to set target values for the SQCs.

We've come a long way in this discussion of QFD, and we've accomplished a lot. What's next?

Many developers have launched into using QFD after studying the parts of the House of Quality as we have just done, only to become mired in details. The key to successful use of QFD is to understand how QFD relates to the organization's overall operations and needs, and to understand how to deal with the practical minute-to-minute, day-to-day details of QFD that drive a team to dispair if not properly managed.

In Part III, we'll look at how the principles of QFD mesh or could be made to mesh with the organization's needs. In Part IV, we'll explore the practical issues of implementing the QFD House of Quality.

Once we've mastered the House of Quality, we'll be ready for Part V, a discussion of the different directions a team can go with QFD after completing the HOQ. Join us on this journey toward being the best in the world!

Discussion Questions

In your development process, how are targets set? Can you link these targets to the competition's performance? to customer needs? Is your organization more likely to be comfortable with numeric or nonnumeric SQCs? If nonnumeric, does subfeature analysis exist to provide a foundation for goal setting? If not, how much work would be required to do the analysis?

Part III

QFD from 10,000 Feet

CHAPTER 12

The Larger Picture: QFD and Its Relationship to the Product Development Cycle

Introduction

In this chapter, we'll examine QFD from the perspective of organizations that develop products and services. These organizations struggle with development issues all the time, and they have developed strategies for dealing with those issues.

Those strategies include creating phased development models; acquiring various development tools, including CAD/CAM hardware and software; and making use of various budgeting and project management tools. Despite these attempts at improving their processes, anyone working within one of these organizations knows that more can be done.

In this chapter, we'll see how QFD can have a positive impact on the development process. First, we'll look at QFD's role in improving communication between development team members. We'll even consider the suggestion that QFD could be the "glue" that helps the team members to work together more efficiently, even if—*especially if*—they come from different disciplines and understand different jargon.

Then we'll look at how QFD's original formulation has been extended ["Enhanced QFD" (EQFD)] to cover some important aspects of development that have always been difficult to systematize: technology and concept selection, in particular.

Finally, we'll look at reliability engineering, and how QFD can provide guidance in this difficult discipline.

12.1   Cross-Functional Communication

For most organizations, the product or service development process was in existence long before QFD came on the scene. The introduction of QFD is often viewed by developers as an add-on—a tool that must or can be used in addition to the existing development processes. An alternative way of viewing QFD is as an organizer, or as the glue that can bind together the many aspects of development.

In Diagram 12-1, the objects arranged in a circle around QFD represent typical organization functions, each of which plays a role in successfully bringing a

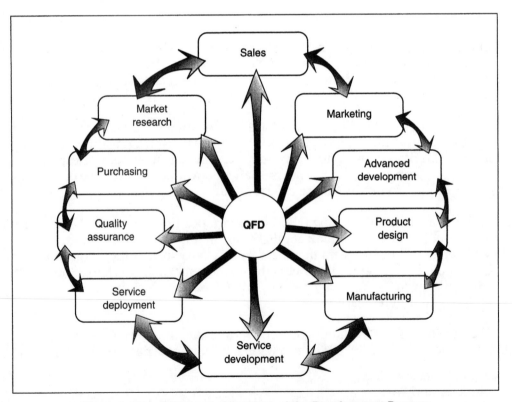

Diagram 12-1. QFD as an Organizer of the Development Process

product to market. In order to execute their functions well, these organizational functions must communicate with each other efficiently, and they must all be focused on a common goal.

For example, Purchasing must negotiate for the best materials, parts, and services, at the best price. The vendor's goods and services must have exactly the characteristics that allow the developers to deliver the product or service that their customers will prefer over the competition. Purchasing cannot work in a vacuum. The purchasing agents must understand what Manufacturing needs and what Product Design has called for, and must understand how these things relate to customer needs. Small misunderstandings can lead to large errors.

Likewise, the Product Design and Manufacturing people must know what's possible in the way of purchasing from suppliers. Are the specifications realistic? Are the right things being specified? Could better things be specified? Are the costs within budgetary constraints? Only a high level of constructive communication among Product Design, Manufacturing, and Purchasing can guarantee the best answers to these questions, and the smartest decisions at the outset of development.

To a large extent, thinking of QFD as being at the center of the communication process helps each functional group to find out how its work fits in, and to tell all the other groups what it needs from them.

Further, putting QFD in the center can enable concurrent engineering, as described in Chapter 2. Imagine the activities depicted in Diagram 2-5 proceeding without a communications tool such as QFD. How could the Manufacturing team and the Design team work toward a common purpose? How could bottlenecks in Delivery development be anticipated and prevented? Only by establishing up-front and ongoing planning that keeps the customer needs in focus and that links actions at every step to these customer needs. If concurrent engineering, or anything resembling it, is to occur in a development organization, QFD will be seen as the principal tool to support it.

Equally important, QFD provides criteria for determining the goodness or appropriateness of any decision. These criteria are derived directly from, or can be clearly traced to, customer needs. Hence, the Voice of the Customer becomes the key backdrop against which communication occurs during the development process.

Development cycles are generally rather long for most products and services. Automobile development cycles in the U.S. were about five years during the eighties, and are now approaching three years during the competitive nineties. Desktop computers take six to twelve months for development by the most competitive U.S. companies (completely new designs take much longer). The time between software releases is generally one to two years. Brand new services from financial companies often take six to eighteen months to launch.

The biggest problem developers face with these long time frames is dealing with the uncertainties of the future. Who can be sure that decisions made on a particular day will be appropriate in the business climate of one, two, or five years in the future? So many things can change during that time. Customer attitudes could change based on political, social, or scientific developments that could not have been predicted. Even if the events could be predicted, the reactions of customers cannot be predicted with certainty.

Against such a backdrop of uncertainty, how can the developer produce the right product or service at the right time? Two strategies, used in combination, can go a long way toward reducing these risks: Reduce cycle time, and stay tuned to the customer. We've discussed the first strategy elsewhere (Section 2.4 and many other parts of this book).

Staying tuned to the customer means developing a clear and detailed understanding of customer needs during the planning stages of development. It also means checking development decisions against up-to-date assessments of customer needs on a continuing basis.

The longer the development cycle, the more likely it is that the marketplace and associated customer needs will change. The change will most likely be a quantitative change—that is, importance levels or satisfaction performance levels on the customer needs may vary. Less likely, but very significant if it happens, is a qualitative change: a new customer attribute may appear. The QFD matrix or matrices can easily be updated with such changes, and development targets and priorities can be reassessed to determine whether the development work as planned is still on track.

Once again, by placing QFD at the center of the organization's development communication model, any changes to the Voice of the Customer can quickly be assessed against each function's activities.

12.2   Enhanced QFD and Concept Selection

When QFD was first introduced into the U.S., the QFD model assumed that the selection of appropriate technology for a product, piece part or service concept was outside the scope of QFD. Don Clausing and Stuart Pugh realized that the process for selecting innovative concepts could and should interact with the translation of customer needs to prioritized technical responses. They embodied their ideas in a process called *Enhanced QFD,* or EQFD for short.[1,2]

EQFD consists of five broad but interrelated processes: Contextual Analysis and Static/Dynamic Status Analysis, Structuring of Product Design Specification, House of Quality, Concept Selection, and Total System/Subsystem Analysis.

12.2.1 Contextual Analysis and Static/Dynamic Status Analysis

Contextual Analysis in EQFD is a comparison of competitive products, for the purpose of deciding at a strategic level how to position the product being planned, with respect to both the business and the technical environment.

In contextual analysis the development team creates a series of pairwise plots of critical functional parameters, across several products. The goal is to characterize the current state of product offerings with respect to critical parameters.

Diagram 12-2 is an example of a single pairwise plot. Gasoline-powered vehicles of a certain type are plotted using two critical performance parameters: vehicle weight versus fuel efficiency as measured by miles per gallon. The plot reveals which vehicles are most fuel efficient (the ones plotted close to the upper diagonal bound), and that all vehicles perform within the range

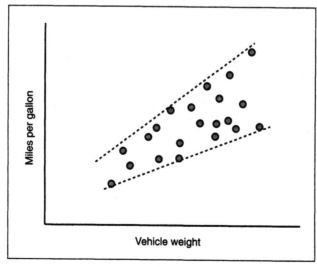

Diagram 12-2. Contextual Analysis—Plot of Critical Parameters

shown by the upper and lower diagonal bounds. Clearly a vehicle that performed lower than the lower bound would be unacceptable in the market (unless it had some major compensating characteristic outside of this analysis), while a vehicle performing above the upper bound would be very competitive.

A wider range between the upper and lower bounds is a possible indicator that the associated technology is *dynamic*, meaning that a relatively wide range of technical solutions exist for achieving the same result. EQFD distinguishes between *dynamic* and *static* concepts.

Static concepts are those that have stopped evolving and have become standardized. The wheel is an example of a static concept—it has been around for thousands of years, and it is almost always used as the locomotor device for vehicles that move over relatively smooth surfaces, such as paved roads or paths. (There are other alternatives, used for specialized situations: runners for ice and snow; treads for uneven surfaces. Most of these are also static for their particular applications.)

The wheel suspension, on the other hand, is considerably more dynamic. Many technical solutions are in use, and for the piece parts of suspensions, a still wider variety of concepts compete with each other.

If a concept is static, the conventional concept is probably the best to adopt. If the concept is dynamic, the team would do well to compare the contending concepts and choose the best for their application, or develop an even better one. The process for evolving to the best solution is called Pugh Concept Selection, described in Section 12.2.4.

In EQFD, developers are urged to perform contextual analysis on critical parameters (as determined by QFD), and static/dynamic analysis on technical concepts. Diagram 12-3 summarizes the relationship between Static/Dynamic analysis and QFD: if the concept is static, first select the concept, then use QFD to determine the requirements. If the concept is dynamic, first use QFD to determine the requirements (using solution-free SQCs), then use the resulting prioritized SQCs and target values as input to concept selection, and select the concept.

12.2.2 Structuring of Product Design Specifications

In EQFD, Clausing and Pugh recommend the use of generic design parameters as a way of beginning the process of developing Substitute Quality Characteristics. Pugh has described these generic parameters in his book *Total Product Design*.[3]

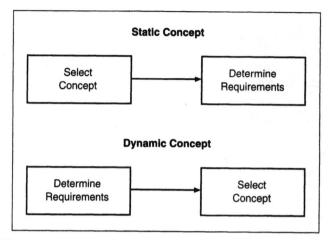

Diagram 12-3. Product Design Based on Static/Dynamic Analysis

The generic parameters include shelf life, packaging, life in service, aesthetics, and more than twenty-five other factors (more than thirty factors in all). As indicated in Chapter 7, this is one of several methods for developing SQCs.

The advantage of using a generic set of SQCs, rather than generating one by the Affinity Diagram method, is that the generic set is likely to include topics that the development team would otherwise overlook. The team can, of course, use or ignore any of the items in the generic list. But by starting with a very complete set of generic topics, they have the opportunity of building their list from a kind of "master" list that represents the best thinking on the topic.

12.2.3   House of Quality

The HOQ in EQFD is the same as in "traditional" QFD.

12.2.4   Concept Selection

The Pugh Concept Selection Process is the centerpiece of EQFD. It helps developers select between alternative concepts, or converge to a concept that's better than the starting alternatives. It helps the developers do this in a way that exploits the best aspects of teamwork, and avoids the worst.

Typically, when developers are confronted with alternative concepts, they align themselves with one or another concept. The process they use is similar to comparing apples to oranges. As the concepts are compared, people tend to focus on one or two strengths, and on one or two weaknesses, of each concept. For example, one automobile design is aerodynamically superior, while another design is less costly to manufacture. This type of unbalanced comparison tends to encourage people to ignore the full complement of criteria that all concepts should be judged by.

When one concept is selected, the other is rejected. The result is that some people "win" and others "lose." The process of concept selection becomes contaminated with this "win/lose" attitude, and emotions dominate the decision process.

There are at least two tragedies associated with the "win/lose" style of selecting from alternative options. The first is that the best aspects of each rejected concept are discarded along with the concept itself. The second is that the "losers" find themselves forced to support the "winning" option. The quality of their support is likely to be quite poor, both because they may not understand the winning concept well, and also because of human nature.

The Pugh Concept Selection Process increases the likelihood that the best aspects of all the alternatives will be reflected in the chosen alternative. This usually means that the chosen alternative will be better than any of the starting alternatives. The Pugh Concept Selection Process also increases the likelihood that the entire team will participate in evolving to the chosen alternative, and will then be more willing and able to support its implementation.

The Concept Selection Process was described by Stuart Pugh as early as 1981.[4] Professor Pugh and others have presented the method to hundreds of developers in the U.S., and its appeal has been immediate.

The general principles governing Concept Selection are
  · Aim for a world-class concept, settle for nothing less
  · Start with the best current concept, even if it's the competition's
  · Either beat the best current concept, or use it

Here's a step-by-step procedure for using the Pugh Concept Selection Process (refer to Diagram 12-4).

  1. Identify the initial alternative concepts. Describe all the concepts at about the same level of detail. Fairly rough descriptions of the concepts will suffice at this stage.

     Represent each concept by a diagram, picture, or cartoon. Sometimes this may be difficult or impossible—for example, with software concepts. In this case, identify key words that stand for the concept. The idea is to

DATUM

	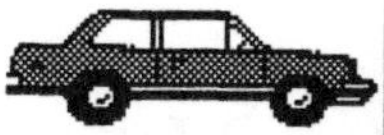	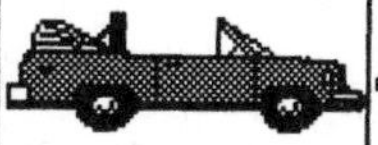		
Response: quicker is better		+	+	+
User maintenance: fewer tasks is better		−	−	−
Ease of mfg.: fewer parts is better		S	−	−
Ease of service: clearer messages is better		+	+	−
Risk: older technology is better		−	−	−
Mfg. risk: more off-the-shelf parts is better		S	−	−
		+ 2	+ 2	+ 1
		S 2	S 0	S 0
		− 2	− 4	− 5

Diagram 12-4. Pugh Concept Selection Process

make the picture suggest as much as possible what is important about the concept at a glance, so that team members can scan the concepts and be rapidly reminded of the essentials of each.

Developers typically start with five to ten concepts—sometimes fewer, rarely more.

2. Develop success criteria. The criteria can be drawn from customer needs data, from product specifications documents, and from business requirements. Express each success criterion along a continuum, so that one of these terms will apply: "Larger the better" (LTB), "Smaller the better" (STB), or "Nominal the best" (NB).

Developers typically generate long lists of criteria. However, it's best to pare this list down to ten to twenty items, to keep the matrix to a manageable size, and because we want the key criteria, rather than all criteria, to drive the selection process. One way of paring the list is to prioritize, using the Analytical Hierarchy Process or the prioritization matrix method. After prioritizing, drop off the least important criteria, leaving the most important ten to twenty items. Another method, if higher level or previous QFDs have already occurred, is to use prioritizations derived from those previous QFDs.

3. Create a large matrix. The physical dimensions of Concept Selection matrices are generally as large as those of QFD matrices: three to six feet high and ten to twenty feet wide.

   Place the criteria along the left edge of the matrix. The criteria should be as legible as possible from a distance. In practice this is hard to accomplish, because the criteria tend to be carefully worded, and are therefore stated in longish sentences.

4. Designate the most promising concept as the *datum,* and place its diagram at the top of the leftmost column. Place the diagrams of the remaining concepts along the top of the matrix, to the right of the datum. The diagrams should be large enough to be seen from the opposite side of the room.

   The datum is the concept to which the others will be compared.

5. Compare each concept to the datum. Start at the top row (corresponding to the first criterion) and ask if the current concept is *better, worse,* or *about the same* in terms of meeting the criterion. If the team's judgment is that the concept is *better,* they enter a "+" in the corresponding matrix cell. If it is *worse,* they enter a "–." If the concepts are *about the same,* they enter a symbol that indicates equality. Stuart Pugh recommends the use of "S" rather than "=," in order to have a symbol that is unlikely to be confused with "–."

   Don't bother to distinguish between "much better" and "a little better." Occasionally measurement may be possible, or existing data may be available. For the most part, however, the team will be using their engineering experience and expertise systematically to make a series of focused consensus judgments. I often remind teams I'm working with that the process is not "rocket science." It's a team method for encouraging creativity tempered by discipline. If the team can develop a concept that *overall,* across all criteria, is better than any concept they started with, they have gotten the primary benefit of the concept selection process.

6. After comparing all available concepts to the datum, the team tallies at the bottom of each column the number of "+," "–," and "S" scores. There are no cookbook rules for developing conclusions based on these scores. The score we are aiming for is one that contains only "+'s"—that is, we want to develop a concept that is superior to the datum across all criteria. There are essentially two possible paths to discovering such a concept.

   The first path is via the brilliant idea, possibly suggested by the discussions and scorings of the other concepts.

The other path is via observing the relative strengths and weaknesses of the concepts, and aiming for a hybrid concept that may be superior to any yet analyzed. In my own experience working with many teams doing concept selection, the "mix and match" idea works well, and often as teams consider swapping features of one concept with another, they get the "aha" that they need for a breakthrough.

7. When the team identifies a concept that is superior to the datum, it's a good idea to promote that concept to be the new datum, take a break from the process, and upon reconvening, generate new concepts that will be better still. There is probably no end to how good the concepts can get; it's limited by the available time a team has for this process. In my experience, teams often work together in concept selection for three to eight long meetings ("long" being a half-day to a day). The additional meetings allow the team to explore fine distinctions between similar concepts, or to branch out to markedly different concepts that may have been suggested by earlier sessions.

The meetings may be held on consecutive days. That is sometimes impractical, however, because of other work some team members may be responsible for, or because the team may need to gather information from other sources before the next meeting.

12.2.5 Total System/Subsystem Analysis

The HOQ is used in EQFD iteratively, first at the system level to develop system-level specifications, then at more detailed levels of design. In Diagram 12-5, we see that the output of the first HOQ is a set of system-level specifications. These become the input to a Pugh Concept Selection matrix that is used to evolve the most appropriate technical concept. This concept, along with the system-level specifications, becomes input to a second QFD matrix (called a "design matrix" in EQFD terminology), the output of which is a set of subsystem specifications. Some of these subsystems may involve dynamic concepts, and are therefore suitable for concept selection, leading to subsystem concepts. These concepts become the basis for lower-level design matrices of piece parts.

Thus, the EQFD formulation provides a structure that product developers can use to guide their technology choices as well as their design. Like QFD, EQFD helps developers reduce development time, increase customer satisfaction, lower product cost, create greater cooperation between functions, establish a better corporate memory, and make fewer mistakes. However, EQFD also provides an explicit road map for the development of complex systems involving multiple levels of design.

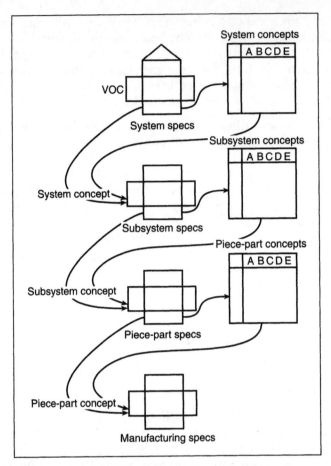

Diagram 12-5. Deployment through the Levels

12.3   Robustness of Product or Service

No product or service development project can escape the necessity of reliability analysis. Some developers plan and design for reliability before launch-
ing their product. Others are forced to do it after product launch, by angry
customers. Although many developers act as if it is cheaper to skimp on reliability design during development, it is almost always an order of magnitude
more expensive, if possible at all, to build in reliability after the product or
service has been launched.

The most common and effective reliability planning techniques are the
Process Decision Program Chart (PDPC), Fault Tree Analysis (FTA), Failure
Mode and Effects Analysis (FMEA), and Robust Engineering.

PDPC[5] refers to the Process Decision Program Chart. This is one of the Seven Management and Planning Tools introduced in Chapter 3. With this method the team draws a diagram—possibly a flow diagram—that describes how a customer would use or interact with their product or service. At each point in the customer's interaction, the team brainstorms all the possible things that could go wrong. For each brainstormed mishap, the team then develops "countermeasures." These countermeasures fall into two categories: actions that can be taken *when* the mishap occurs, and—better—actions the development team can take to *prevent* the mishap. PDPC diagrams can get huge; several methods exist to focus the team on the most likely and the most serious possible mishaps.

Fault Tree Analysis[6] refers to the systematic decomposition of possible product or service failures into the component causes, based on the *function* of the product or service. Analysis usually begins with a list of the functions, possibly generated from a functional tree. For each functional failure, several causes can be generated by the development team. Each of these causes will in turn likely have several causes. Thus, the development team can build a complete Fault Tree. Having completed this analysis, the team must then decide how to best design the product or service to prevent the failures.

Failure Mode And Effects Analysis[7] refers to the systematic analysis of possible product service failures, based on the *design* of the product or service. Analysis begins with the major components of the product or service. For each top-level component (or subsystem), the team identifies the possible ways that component could fail. The analysis can extend to components within each subsystem, as well as to the subcomponents and parts. Each failure mode is described in terms of the nature of the failure, the possible causes of the failure, and the effect and impact of the failure. In addition, the team estimates for each failure mode the probability of its occurrence, and the severity of its effect. Sometimes teams also document safety issues, repair or recovery processes, and plans for preventing and/or responding to the failure.

Robust Engineering[8] is an extension of the statistical field of Experimental Design. This extension, originally developed by Dr. Genichi Taguchi, and now practiced increasingly in the U.S., helps engineers identify those design factors that have the dominant impact on the performance of their system. It further helps them select target values for these dominant factors such that the system performance will be robust—will operate consistently—in the face of environmental variation, manufacturing variation, and product deterioration over time.

Detailed discussion of these techniques is beyond the scope of this book.

For each of these methods, the development team is faced with an imposing list of areas where they could devote time and money to improve the reliability or robustness of the product or service. For almost all development projects, it is impossible to perform all the reliability work they have identified. They must select some areas to work on, and leave other areas alone. (Naturally all life-threatening aspects of the design must be attended to.) Once the reliability analysis is complete and all possible reliability areas have been identified, the team can then turn to QFD to guide them in prioritizing which areas to work on.

Summary

In this chapter we have examined the impact QFD can have on the overall development process. Its impact spans three major areas: team members' communication; their coordination; and reliability engineering.

Regarding communication and coordination of team members: QFD's power lies in its relationship to the entire product or process development cycle. It should be positioned as the primary communication tool, facilitating and even defining communication between functional organizations.

With regard to technology and concept selection: Enhanced QFD (EQFD) is a product development schema having QFD at its center, but also incorporating other key ideas. Thus, it further develops the notion of QFD's influence on the development process. Besides the HOQ, EQFD embodies Contextual Analysis, Static/Dynamic Analysis, Concept Selection, and a Total System/Subsystem deployment through the levels.

Finally, with regard to reliability engineering: a key activity in any development activity is designing for reliability. While powerful techniques such as PDPC, Fault Tree Analysis, Failure Mode and Effects Analysis, and Robust Engineering are available, by themselves these methods cannot guide developers as to which aspects of the design most need reliability engineering. QFD as a prioritization tool plays an important role in helping to focus reliability engineering resources on the critical reliability issues.

Thus, we see how QFD can make the development world a better place to be for developers and their customers. Unfortunately, there are many obstacles along the road to this better place. These obstacles are the same ones that make it so hard for organizations to effectively adopt Total Quality Management or any of its aliases.

Given these obstacles in *our* organization, should we give up on QFD? This author recommends no such thing. QFD will have a reduced positive effect in an imperfect environment, but the effect will still be positive. In the next chapter, we'll discuss what it might be like to use QFD under less than ideal circumstances.

Discussion Questions

What product development information gets communicated across functional lines? What mechanisms support the communication in your organization? Which functional groups regard themselves as the least influential? How might QFD help them to change their self-assessment? What benefits might accrue if these groups regarded themselves as more influential in product development decisions?

How does the Advanced Development function decide which technologies to invest in? What activities do they pursue that resemble Contextual Analysis and Static/Dynamic Analysis? Classify the key concepts in your product or service according to the Static/Dynamic dimension. Share your results with your colleagues. Do they agree with your analysis? Given your analysis, how do concept selection activities in your organization match up? What are the advantages and disadvantages of your organization's concept selection activities versus Pugh Concept Selection?

How is reliability engineering done in your organization? Compare your organization's tools with PDPC, Fault Tree Analysis, and Robust Engineering. How are priorities set? How might QFD change priority setting for reliability engineering?

---

[1] "Enhanced Quality Function Deployment," series of five instructional videotapes, MIT Center for Advanced Engineering Study (no copyright date appears on these tapes).
[2] "Enhanced Quality Function Deployment," Don Clausing and Stuart Pugh, Proceedings of the Design and Productivity International Conference, Honolulu, Hawaii, February 6–8, 1991.
[3] *Total Product Design*, Stuart Pugh, Addison-Wesley, Reading, Massachusetts.
[4] S. Pugh, "Concept Selection—A Method That Works," Proc. I.C.E.D. Rome (March 1981), WDK 5 Paper M3/16, pp. 497–506.
[5] "Memory Jogger Plus+II," GOAL/QPC.
[6] *Total Quality Development*, Don Clausing.

[7] *Quality Planning and Analysis*, J. M. Juran and Frank M. Gryna, Jr., McGraw-Hill, New York, 1980.

[8] *System of Experimental Design*, Genichi Taguchi (Don Clausing, technical editor for the English edition), UNIPUB/Kraus International Publications, 1987 (2 vols.). See also Phadke, *Quality Engineering Using Robust Design*.

CHAPTER 13

QFD in an Imperfect World

Introduction

This chapter covers the real-world scenarios of corporate life, where the customer is *not* king, and where quality processes are *not* uppermost in every employee's mind.

In this chapter, we'll look at scenarios where some team members resist QFD. How can the visionary manager or QFD facilitator help people to benefit from QFD in the face of opposition? While the topic of organizational change is beyond the scope of this book, anyone attempting to introduce QFD into an organization will do better by anticipating the possibility of resistance.

Much of the resistance to QFD comes from its multifunctional nature. Because it helps team members from different functional groups to communicate, it appears to some members of these groups as if the other team members are taking over their job. In this chapter, we'll explore these "turf" issues, and look for constructive ways of introducing QFD.

The QFD process—especially the planning of QFD, acquiring the VOC, and developing the House of Quality—has the effect of broadening each person's concept of what a successful product or service must consist of. Some product developers welcome the implied broadening of their responsibilities; others don't. Welcome or not, however, the broadening effect exists.

Diagram 13-1 shows a simplified view of processes within a corporation that relate to developing products and services. As with many organization charts, this diagram does not contain horizontal arrows or lines to indicate communication and decision-making that cross, or that should cross, organizational boundaries. In the ideal organization, the organization's infrastructure would support such cross-functional communication and decision-making, and QFD would be part of that support.

In many situations, however, the organization has not evolved to the point where cross-functional communication is easy and natural. QFD, if it exists, is used in isolated areas, and therefore its full benefits are not gained. For many QFD advocates, this is a fact of life. All is not lost: the full set of possible benefits is not available, but some benefits may be. It's a good idea for the realistic manager planning to use QFD to understand the possible

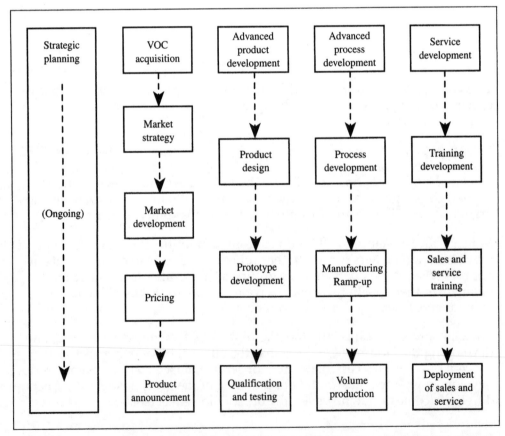

Diagram 13-1. Traditional Uncoordinated Corporate Product Development Processes

"partial-benefit" scenarios. Let's take a look at some of the most common of these scenarios.

13.1   Marketing Functions in Engineering-Driven Environments

Marketing departments react to QFD in various ways, depending on the organizational culture.

In technically driven organizations, product definition emanates from engineers and scientists. Hence, the activities of the marketing department are normally centered on overall sales strategy and management of key accounts, rather than on product definition.

QFD is often viewed with suspicion by people in such marketing roles, because the QFD paradigm explicitly suggests that product definition be shared with other organizations, in particular with the technical organizations, which actually perform the function. In other words, QFD may be seen as a threat to turf that marketing claims but is insecure about.

The rational marketing professional's response to QFD in this case might be: "QFD provides me a way of having more influence on the product development process than I currently have." Should a QFD "champion" encounter resistance to QFD by marketing departments, very likely fears of losing control over coveted turf are at issue.

13.2   Engineering Functions in Marketing-Driven Environments

In marketing-driven organizations, it may be the engineers or the technical establishment that is suspicious of QFD. There may be technical leaders, accustomed to the role of approving or dashing to the rocks project proposals that emanate from the marketing organization. Typically these technical gurus retain control of murky technologies that no one else in the organization understands. They tend to be unfamiliar with customers or with the wide range of customer needs that bear little relationship to the technical aspects of the product.

The rational technical leader's response to QFD might be: "QFD provides me with a way of off-loading the uninteresting nontechnical aspects of product definition on others who can combine their knowledge of the customer with my knowledge of the technology."

13.3   QFD in Engineering-Driven Organizations

Engineering-driven organizations generally develop products that are centered around advanced and dynamic technical concepts. Examples are

- Computers that are faster and cheaper than previous models, because of advanced semiconductor technology
- Tennis racquets more lightweight than previous models, because of advanced materials development
- Medical instruments that permit surgery without cutting open large areas of the patient's skin, based on breakthroughs in the combined areas of medicine, optics, and micromechanical engineering

These products may have mixed success in the marketplace. Sometimes the dominant technology fills customer needs, as with lightweight tennis racquets. In other cases, the dominant technology is irrelevant to market needs, as with mainframe computers in the early nineties, with demand continually growing for workstations and desktop computers.

The typical problem in engineering-driven organizations is that engineers, for one of several reasons, don't listen very much to the Voice of the Customer.

The most common reasons are these:

- Engineers believe they already know what customers want.
- Engineers don't regard listening to customers as "part of the job."

A common approach to QFD in organizations like this is for engineers to consider QFD a high-level design tool, the use of which need not affect other functions in the organization. Many engineering groups have developed the affinity diagram of customer needs according to their own belief of what the customers want. They have then gone on to estimate customer importance of needs and customer satisfaction, usually by consensus of the engineering team, and have developed the entire House of Quality on their own (see Diagram 13-2).

This approach would seem to sidestep the primary purpose of QFD, considering QFD as the tool that is meant to drive the Voice of the Customer into product design. Indeed, as we have seen (with the U.S. auto manufacturers in the late seventies and early eighties, for example), many large companies have suffered greatly by not responding to their customers.

Despite this misapplication of QFD, there are still some benefits to building a House of Quality, however incorrect the basic inputs to it are. The main benefits are these:

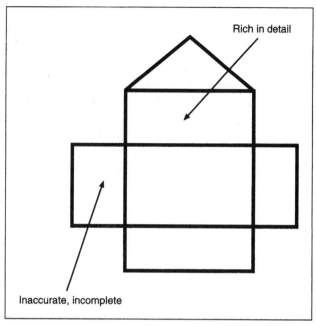

Diagram 13-2. Engineering-Driven QFD

- The group's attempt to create an affinity diagram of all customer needs usually results in a far broader, better-structured description of the customer than any single group member could have developed alone.

- The entire team gains a shared, even if incorrect, vision of the customer's needs (and how the Substitute Quality Characteristics relate to them). Most other methods of top-level product specification begin with an elite few engineers or scientists. These technocrats conceive of a product concept while closeted away in a private area. Once they have worked out their ideas, they announce their concept to the rest of the team, and attempt to win them over. A common alternative to this approach involves some iterated series of presentations of their ideas to management and to the rest of the team, in which they accept comments and objections, and respond to them in subsequent versions of their ideas.

  Whether the remainder of the team eventually comes to understand their vision, or agrees with it and is willing to support it, is a function of the leadership and the persuasive skills of the original elite designers. This "selling of the product vision" is a hit-or-miss affair, in which the elite designers "win" if their idea is accepted, and "lose" if it is not. Conversely, those who must be sold often resist the new idea in order to avoid conceding a "win" to the elite engineers, since conceding a "win" implies they must have "lost." Of course, it's the organization that suffers the greatest loss, since this selling process generates imperfectly understood concepts, along with winners and losers. No organization needs imperfectly understood concepts, and no organization needs losers.

By contrast, with the QFD process a much larger team participates in developing the concept from the start. They work together on it, and emerge with a common understanding of it. They jointly develop a detailed understanding of the relationship between their understanding of customer needs and of corresponding Substitute Quality Characteristics. The customer needs may be inaccurate, but at least the whole team can support the project. A product developed by a team with a common vision is more likely to succeed than is a product developed by a divided team.

Thus we see that even when some of the basic principles that QFD supports (starting with the correct VOC, and using cross-functional teams) are violated, the use of QFD can have some benefits. Furthermore, when teams *do* violate basic principles in the use of QFD, their errors become visible to them, thus creating motivation to do the job right the next time.

13.4   QFD in Marketing-Driven Organizations

Marketing departments in strongly marketing-driven organizations tend to regard downstream activities such as purchasing and manufacturing as irrelevant to product planning.

In these environments, QFD activities often involve the Market Research and Product Marketing functions with little or no representation from Engineering and Manufacturing.

The QFD activity may be strongly driven by the Voice of the Customer, but because of technical under-representation may also poorly reflect technical and implementation aspects of the product or service. The result, as with the Engineering-Driven QFD, will be a product planning process that is off the mark (Diagram 13-3). Nevertheless, some benefits are still available to the participants: common understanding of the customer and of a desired direction, as well as increased ability of the participants to communicate with each other.

13.5   Manufacturing and QFD

The manufacturing organization can benefit from QFD, even when it is used in isolation, in much the same way that engineering-driven QFD can provide benefits to the engineering team.

The manufacturing manager or manufacturing engineer needs to know how the manufacturing function's work affects customer satisfaction. It pays to

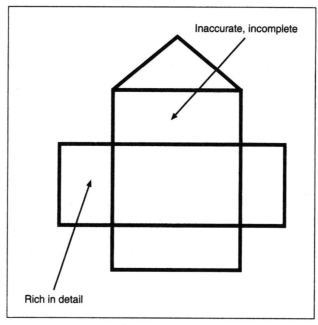

Inaccurate, incomplete

Rich in detail

Diagram 13-3. Marketing-Driven QFD

create a QFD matrix in which the Whats are expressed as either the Voice of the Customer or a derived language linked to the VOC.

The Hows can be used to describe manufacturing's activities expressed in a suitable language. As with every other part of QFD, it's best if the Hows are expressed as performance measures.

Sometimes when manufacturing works in isolation, a matrix will be constructed that reflects that isolation. Such a matrix may have a piece-part specification for the "Whats" and process parameters for the "Hows," as in Diagram 13-4.

Manufacturing organizations are not alone in having multiple customers and suppliers. Possible partners with the manufacturing organization include those functions shown in Diagram 13-5. In some cases the partner is primarily a customer, in others primarily a supplier; in still other cases, the partner is both a supplier and a customer.

For each relationship with another group in which manufacturing is either a customer or a supplier, the What versus How analysis will lead to a better understanding of how the relationship should work.

The standard QFD models indicate that the Voice of the Customer can and should be deployed all the way from high-level system specification down to

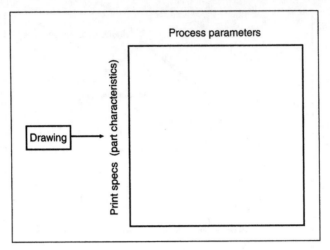

Diagram 13-4. Piece-Part QFD When Manufacturing Is Isolated

the factory floor. As of 1991, survey data indicated that fewer than one QFD application in five used QFD for production.[1] There is no evidence in recent QFD literature that the ratio has changed overall, although for certain organizations (such as Chrysler Corporation) QFD is being used increasingly down to the shop-floor activities.[2]

13.6   Sales and QFD

Besides the potential supplier/customer relationships that every sales organization has with groups internal and external to its own corporation, and the benefits possible from What versus How analysis, the Sales organization may have a particularly interesting opportunity for using QFD if its business involves generating sales proposals of any complexity as a normal part of business.

Some sales teams have successfully collaborated with their customers to create a set of customer requirements (the "Whats"), and have used QFD to demonstrate how their proposal (the "Hows") met their customer's needs.

13.7   Service and QFD

We've already seen several examples of QFD used for planning service activities. Internal service groups, such as Human Resources and Site Management groups, have gained insight into their functions by defining their internal customers, interviewing their customers to determine the attributes of customer satisfaction, and then evaluating their existing programs against these internal customer wants and needs.

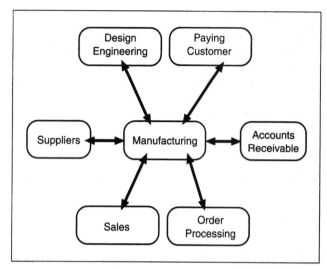

Diagram 13-5. Manufacturing, Its Customers, and Suppliers

13.8   Exerting Influence beyond Organizational Barriers

When product developers embark on a QFD project, their first step is to acquire the structured customer wants and needs, the Voice of the Customer. This process often makes developers, especially design engineers, uncomfortable, because the Voice of the Customer tends to include elements that the developers feel they have no control over. When collecting the Voice of the Customer, developers often learn that customers care about "easier order processing," "on-time delivery," or "defect-free manufacturing," topics that product engineers typically have little experience with or influence upon. Listening to the voice of the customer often has this effect of widening the scope of the problem to cover functions and skills that are not currently involved. What is a product developer, or any other QFD team member, to do with such customer needs?

The answer obviously varies with the organization. In some cases, there is simply no one available to listen to and respond to these needs. The QFD team may not be truly cross-functional; for example, there may be no one present to represent the order-processing function, so the team may feel powerless to improve the order-processing system. This is an example of how important it is to develop products and services with cross-functional teams that truly cover all the functions. If each functional group really has an impact on the customer's perception of the product, then each group's inputs into the product from the very beginning will have the greatest positive effect on the end result.

Let's take customer dissatisfaction with order processing as an example. Order processing may be seen by product developers as a fixed, unalterable process of great weight and complexity. Its vagaries may be regarded as difficulties that everyone, product developers and customers alike, must put up with. This attitude, that a process owned by another department cannot be altered, is of course a symptom of the departmental barriers that build up in larger organizations. These barriers lead to inflexibility and paralysis in organizations and ultimately make them unable to compete. They are the very barriers that QFD can help break down.

It doesn't have to be this way. Product development teams can actually influence such processes as order processing in at least two ways, far beyond what they may believe. The adage "fight 'em or join 'em" provides a useful way of looking at how product developers can deal with the problem.

The *Fight 'Em* approach aims at forcing or inducing an improvement to the order-processing function. The *Join 'Em* approach aims at accepting the order-processing function as is, and at making the product robust against the process's weaknesses.

The Fight 'Em Approach. This usually means motivating ("It's the right thing to do") or forcing ("Do it or else") the owners of the order-processing function to fix their system.

Motivating often starts with the Voice of the Customer. The developers can confront the order-processing system owners with their VOC findings. Assuming the data is credible, the reasonable, nondefensive order-processing system owners are likely to want to change their process to better meet the needs of their customers. They can (and should) use QFD and concept selection to develop the desired process changes. If the order-processing system owners are so inclined, they can align their QFD activities with the product development activities, and begin breaking down barriers between departments.

Forcing changes generally involves going over the heads of the order-processing owners by revealing the VOC data to levels of management that can mandate order-processing changes. Forcing changes in this way is obviously a method of last resort, but the author has seen such methods work, after other attempts have failed. Forcing changes is most effective when the VOC data is credible, easy to understand, and kindles an emotional response (see "Representing the Voice of the Customer").

The Join 'Em Approach. This means adapting the product design so that it avoids or sidesteps the weaknesses of the order-processing system. The most common situation relates to products that can be customized to many configurations, such as automobiles, computers, or office panel systems. These products come with many options. Some options are independent of all other options, as for example tinted windows in automobiles: these windows could be made available for all car models, and would have no mechanical dependence or interdependence with other options. Other options won't work or are otherwise dependent on the existence of yet other options in the product—for example, airbags, which may be installable in certain steering columns and not in others.

The existence of many options, with a multitude of interdependencies, creates untold difficulties for everyone and every process connected with order processing. Sales people and order takers must be aware of all the complexities and must be able to explain them to customers, who must be capable of absorbing them and evaluating and selecting the options that are right for them. Everyone involved must possess a vocabulary to formulate each option, and to understand the differences between the many configurations.

Pricing is, of course, a major problem. Besides discount structures, which are normally customer-dependent, the prices for various options and configurations of a product can be difficult to compute and justify.

To design a product to be robust with respect to an order-processing system that cannot easily handle these complexities is to design a product that is simple to understand, has relatively few options, and yet satisfies a wide range of customers. Here is a challenge that QFD can help teams to understand.

Listening to the Voice of the Customer can be disturbing to design engineers, because they will be exposed to aspects of the product that are beyond their normal view of what they are responsible for. This is exactly why QFD can be valuable to a development team. It forces developers to redefine and broaden their scope, and it helps them to design products that more fully satisfy customers. In the end, these are the products that will keep their company competitive.

Summary

In this chapter we have seen that although the ideal environment for successful implementation of QFD may not exist, teams can still benefit from QFD. The principle of linking a team's actions to its customers' needs is a worthwhile one to support under any circumstances. While market-driven organizations differ substantially from engineering- or technically-driven organizations, the common element between them is that one function or the other may hold power in product development planning that it is reluctant to give up to a cross-functional QFD team.

The discussion in this chapter has shown that even if this obstacle cannot be overcome, teams can use QFD within their own domain, using the best possible data available to them. The difference in the results of using QFD in this limited fashion will stand out in stark relief against the ideal situation. This difference can be exploited to move the organization in the right direction.

In cases where top-down leadership lacks the vision of the customer-driven organization, strong middle managers have introduced QFD. They have used QFD to highlight the lack of good customer information as well as the need for it. QFD can be infectious in such organizations. As it grows in popularity it can back top management into an uncomfortable corner from which the only way out is to acknowledge the organization's dependence on the customer.

This completes our survey of issues relating to the corporate environment of QFD. We have looked at the need to create efficient horizontal as well as vertical communication in the organization, and at an important enhancement to QFD that provides a framework for managing technology within the context of meeting customer needs. We have also acknowledged that even in less-than-ideal circumstances, QFD can play a vital role.

Now that we've had a look at the trees and the forest, it's time to "walk the talk" by taking a hike along the QFD implementation trail. The next section of the book deals with QFD implementation.

Discussion Questions

Compare your organization to the profiles presented in this chapter. Which profile is closest, and which is farthest from your situation? Share your analysis with colleagues. Do they agree? If not, what are the differences between their analysis and yours, and what accounts for the differences?

How far from the ideal multifunctional team model will you have to diverge in order to implement any kind of QFD project in your organization? What would be a fruitful objective for your organization's first QFD?

---

[1] "Quality Function Deployment: A Study of Implementation and Enhancements," Amitabh Pandey, MIT Master's Thesis, 1992.

[2] "QFD Implementation at Chrysler—the First Seven Years," Robert Dika, Chrysler Corporation, Transactions from the Fifth Symposium on Quality Function Deployment, 1993.

PART IV

QFD Handbook

CHAPTER 14

Introduction to the Handbook

Understanding the QFD matrices is not the same as doing QFD. Most people, as they participate in their first QFD activity, get confused. They usually don't know how long the process should take, or what to do next. The process is complex enough so that people get lost among the trees and have difficulty finding their way out of the forest.

Other parts of this book have described what QFD is, and how it fits in. The QFD Handbook focuses on how to implement QFD, with special emphasis on the House of Quality. If a development team can gain a level of confidence with the HOQ, the various possible follow-on matrices will be relatively simple to develop.

While QFD has been associated with many successful development activities, there are also many examples of QFD projects that have gotten bogged down and eventually abandoned. Successful QFD projects depend on many factors. Some of these are

- Management support, fueled by faith in the benefits that QFD promises
- Complete, credible customer data
- The right development team
- Thorough planning of the QFD project
- A skilled neutral facilitator

In the QFD Handbook, we'll expand on these and other factors. The Handbook is intended to show you the range of choices available for structuring a QFD activity. It is also intended to make the QFD process as real as possible for you, so that you can anticipate the many issues and obstacles that may come up along the way.

Before you start your first QFD project, read this handbook. Then imagine going through the phases described here, one by one. Try to imagine how each team member will react to the work of each phase. Who will support the effort? Who will oppose it? Who will slow things down? Which aspects of the process will be familiar to the team? Which will seem strange and therefore a bit disorienting?

As you review these and other possible obstacles, try to imagine how you will deal with them. Your role could be facilitator, manager of the development group, or member of the QFD team. Whatever role you play, by anticipating what may go wrong in the process, and by visualizing what you'd like the process to be, you will be able to influence it constructively.

Just plunging in to QFD is probably a formula for failure. Careful planning and time management are major predictors of success. The QFD Handbook will help you plan and manage the work.

In the Handbook, QFD will be presented as a *phased activity*. The purpose of describing it this way is twofold. First, the phases refer to distinct types of activities that should be performed more or less in sequence in order to make the QFD effort as efficient as possible. Second, the phases provide guideposts to help ground the participants as they work their way through QFD.

We've numbered the phases from 0 to 3. Phase 0 is the planning phase. Because the work of QFD hasn't started yet, it can be thought of as a "prephase"; hence the number 0. The other phases are

    Phase 1: Gather the Voice of the Customer
    Phase 2: Build the House of Quality
    Phase 3: Analyze and Interpret the Results

Subsequent phases of QFD—deployment of SQCs to processes, for example—can be thought of as additional phased QFD projects, each with its own plan and implementation.

An important success factor for QFD is the support of management. Besides the support of top management, which cannot be underestimated, the actions and attitudes of the QFD manager are critical. Let's take a look at who the QFD manager is, and what the QFD manager's role should be.

14.1   The QFD Manager's Role

QFD is a complex activity, involving the coordinated efforts of several people. In other words, QFD should be treated as a project. We'll talk about making time estimates and setting up schedules later, but the key point for now is: The QFD effort needs a manager. That manager could be the QFD facilitator (see Chapter 17 for the QFD facilitator's role)—the impartial person who runs the QFD meetings. Or it could be one of the team members whose job is closely linked to the success of the development project, such as the marketing manager or the development manager. Whoever the manager is, that person is responsible for the success of the QFD activity and would do well to manage it with the same professionalism and discipline that would be brought to bear on any other project.

Here are some "do's and don'ts" for QFD Managers.

DO:
- Make sure you and your team understand and agree on the benefits you wish to receive from the QFD activity.
- Line up the people who will be needed for QFD. Make sure they know when and for how long they will be needed.
- Establish a schedule for each of the QFD phases.
- Track progress of the QFD activity, and continually seek ways to keep the activity on schedule.
- Create a mechanism for keeping people in your organization who are not on the QFD team up to date. Ensure that their concerns and ideas are represented on the QFD team.
- Ensure that the QFD team members have been brought "up to speed" in terms of their knowledge of the development project, of QFD, and of the QFD project.
- Make use of time between meetings by assigning data gathering and other research-type functions to the QFD team members.

DON'T:
- Drive the QFD project to a foreordained conclusion. If you cannot keep an open mind as to the outcome of the QFD process, it would be better to simply announce your desired conclusion and skip the QFD.
- Assume that everyone will know what to do. Instead, explain what's going to happen in each meeting, well before the meeting day, and at the beginning of each meeting as well.
- Allow the QFD team to make decisions without data. At the very least, point out to the team that they are doing so when you see it happen. Obviously, some development decisions do get made with no data—it

can't be avoided completely. The QFD process can be used to make such decisions more visible. This gives the development team the opportunity to consciously decide whether they really want to take such a risk.

Summary

We've got our project, we've decided to do world-class development (and therefore we're going to use QFD), and we've got our QFD team and its manager. We're ready to launch the QFD activity. Let's start by planning the QFD work.

CHAPTER 15

Phase 0: Planning QFD

Introduction

This chapter deals with the critical process of planning the QFD activity. In this chapter, we'll break planning down into several key planning topics, and we'll explore each one in detail. The key topics are

- Establish organizational support
- Determine objectives
- Decide on the customer
- Decide on the time horizon
- Decide on the product/service concept
- Decide on the team
- Create the QFD schedule
- Acquire the facilities and materials

This chapter assumes you know what QFD is, and that you have familiarized yourself with the ideas covered earlier in this book. This is a long chapter, because planning—the subject of the chapter—cannot be completely separated from doing. As a result, we'll be discussing a lot of QFD implementation steps along with the planning of those steps.

Planning the QFD project is the key to success later on. Developing the House of Quality typically involves considerable expense for gathering the Voice of the Customer (twenty thousand to several hundred thousand dollars for commercial-volume products and services), and then involves a team of eight to fifteen people working together in a room for two to ten days (16 to 150 person-days). Why not develop a plan to ensure the success of such a resource-laden activity?

One of the QFD facilitator's jobs is to see to it that the activity is properly planned. This chapter identifies the major areas that must be planned, and the questions that must be answered.

15.1   Establish Organizational Support

Organizational support for QFD is a key determiner of success.[1] The key elements of organizational support are

- Management support
- Functional support
- QFD technical support

Management support refers to a commitment by top management in the organization to provide and allocate the resources needed to complete the QFD activity. The resources include whatever time and money it takes to gather the voice of the customer, whatever it takes to acquire the services of a skilled QFD facilitator, and whatever it takes to keep the QFD team focused on QFD until the desired results have been achieved.

Functional support refers to the commitment of related functional groups to participate in the QFD activities as needed, and to honor the decisions of the QFD team during the subsequent development process. The functional groups referred to here are those that will be partners with the development team in completing the development of the product or service. For product development, these functional groups could include Purchasing, Manufacturing, Quality Assurance, Sales, and Service. For process development, these functional groups could include Purchasing, Training, Marketing, and Finance.

QFD technical support refers to the acquisition of skills necessary to implement the QFD. Everyone on the QFD team will require at least an acquaintance with QFD principles, preferably through a short training seminar, although many team members will be able to learn what they need to know as the QFD process unfolds. The QFD facilitator will need to be quite familiar with the various elements and options of QFD, in order to guide the team into using those parts of QFD that will help them achieve their objectives. The QFD facilitator will also need to have good group facilitation skills in order to help the QFD team achieve its goals and manage its time well.

15.2   Determine Objectives

QFD provides an array of possible benefits to the teams that use it. Many development managers imagine that the benefits of the process are self-evident and need not be articulated or selected. Quite possibly these managers have simply focused on those few of the many benefits that matter to them, and have overlooked the others. The facilitator should present the development manager with a choice of the possible benefits and ask him or her to specifically identify the ones that apply to the upcoming project. What follows is a list of possible benefits:

- Understand customer wants and needs.
- Determine quality and business goals for the product/service.
- Rank-order proposed product capabilities.
- Develop a common team vision of the product/service.
- Document all decisions and assumptions about the project in a single, compact diagram (the House of Quality).
- Create a list of actions that will move the project forward.
- Develop clear linkages between technical decisions and customer needs.
- Minimize the risk of midproject restarts. This benefit accrues because new information made available in the middle of development can be put in perspective by adding it to the existing HOQ and other QFD matrices. Without such a discipline, new customer information or new technical information is often given disproportionate emphasis, leading to overreaction.
- Rapid product planning. Although QFD appears to be time-consuming, most groups find that product planning is quicker, more complete, and more efficient when they use the House of Quality structure. This is because QFD provides a structure that can be managed and planned for, in contrast to most product-planning activities nowadays.

All of these benefits have been discussed elsewhere in this book. By knowing in advance which benefits are most important to the development manager, the facilitator will be able to allocate more time and attention to those parts of the QFD process that support those benefits. Clearly stating the objectives to the entire team helps everyone to use the process to best support those objectives.

15.3   Decide on the Customer

15.3.1   The Importance of Clear Definitions

During the QFD process, the team will be making many judgments. They will be estimating the relationships between product or service capabilities and customer needs, for instance. In order to make these judgments meaningfully, the team will need to make clear and consistent definitions.

To the extent that a product capability or performance measurement, for example, is vague, the team runs the risk of tacitly defining it one way in the morning, another way in the afternoon. Sometimes during a discussion, team members will discover their disagreement is because some team members are assuming one definition, while others are assuming another.

The resulting values the team puts into a quality table will reflect one view from the morning exercise, another from work in the afternoon. If the team is not clear about its underlying assumptions as it builds the quality table, the results will be inconsistent and will have little value.

The team's most important underlying assumptions will be those about the customer. In my experience, it's surprisingly difficult for product development teams to agree on who their customer is. QFD becomes a magnifying glass of disagreements among team members. The most numerous disagreements tend to be about customers' needs, first because team members don't really know their customers well, and second because different team members tend to focus on very different customer types, without realizing it.

Some people assume the customer is the person who makes the decision to buy the product or service. Others assume it's the person who actually uses the product. These may be very different people. For example, the buyer of a toy is usually the parent, but the user of the toy is usually a child. In a more complicated situation, the buyer of a fleet of fifty delivery trucks may be a fleet manager, but the users of the trucks may be fifty drivers. In addition, five mechanics may maintain these trucks. Who is the customer?

Another complication lies in market segment definition. Returning to our toy example, is it intended for boys or girls? What age group? How much education? What type of physical dexterity? And for our trucks, will they be driven by the same person each day, or will they be rented by the day to different drivers? What type of cargo will they be carrying? Will they be used in the city or the country?

Obviously, if the team has not discussed and come to consensus on these questions, team members will be talking at cross purposes as they proceed through the QFD process. If the team *has* discussed and come to consensus, they will work more efficiently and harmoniously with each other.

15.3.2 Identify All Possible Customers

The first step in defining the key customer is to make a list of all possible candidates. This is usually done by the QFD planners or the market research experts. The affinity diagram is a useful tool for managing this list of customers.

Start by brainstorming all possible customers of the product or service you are planning. Brainstorm onto Post-it Notes. Wherever possible, be specific. Use the names of actual customers, or even the names of people. When mentioning companies, company types, or other organizations, try to replace these abstract entities with people or roles within the organizations. Remember, the quest for customer satisfaction is a quest for satisfying people, not organizations.

Use the affinity diagram method to group the brainstormed items. General categories of markets, user types, or product application types may emerge. From these categories, work toward a list of clearly defined customer groups. One test for clear group definitions is to identify a handful of actual customers and see if all team members place them in the same groups.

15.3.3 Identify the Key Customers

Now that we have identified several customer groups, we must focus in on the key customers. The idea is to optimize our product design decisions around these key customers, and then try to include as many additional customers as possible in our plans. Here are three ways the team might decide on the key customers:

- Everyone agrees quickly
- Prioritization matrix method
- Analytical Hierarchy Process[2]

Everyone Agrees Quickly

Once the customer groups have been identified, deciding on the key customers is sometimes easy.

Everyone glances at the list of customer groups and with little or no disagreement, they are able to decide who are the key customers.

More often, the number of customer groups may be pretty large. Or, the types of groupings may not be comparable (for example, European banks vs. information systems managers at banks). Under these circumstances, teams typically have considerable difficulty coming to consensus on the key customer group.

If everyone cannot quickly agree on the key customer group, one of the other methods for selecting the key customer group may be useful.

Prioritization Matrix Method

The Prioritization Matrix method uses a technique similar to the weighting and multiplication of columns in the QFD House of Quality Planning Matrix.

We identify one, two, or (rarely) more factors or criteria by which we will rate all the customer groups. The following might be typical factors:

- Revenue potential of this customer group over the next three years

    High revenue potential
    Moderate revenue potential
    Low revenue potential[3]

- Current revenue derived from this customer group over the past three years

    High revenue derived
    Moderate revenue derived
    Low revenue derived

- Sales force familiar with this customer group

    High: Most of the sales force familiar
    Moderate: Some of the sales force familiar
    Low: Almost none of the sales force familiar

For each customer group, we then evaluate that customer's importance based on the high/moderate/low judgment in the dimensions we have chosen. We assign numerical values as follows: high = 3, moderate = 2, low = 1.

For each customer group we multiply the numerical values for all factors we have assigned. The resulting product is an indicator of how important the customer group is to us (Diagram 15-1).

	Revenues potential	Recent revenue	Importance
Segment A	3	2	6
Segment B	1	1	1
Segment C	2	1	2
Segment D	3	3	9

Diagram 15-1. Prioritization Matrix

If we have evaluated customer groups according to two factors, as is most common, possible products of factors are 9, 6, 4, 3, 2, and 1. The customer groups with factor products of 9 and 6 are our key customers.

Some refinements of this method are occasionally used: to treat one criterion (column) as more important than another, increase the range of numerical impact values. Instead of using 3, 2, and 1, for example, use 5, 3, and 1. Some prefer to add the impact values rather than multiply them. Multiplying creates greater differentiation in the importance column; adding creates less. Some teams use the "spend-a-buck" method in each column. They require that the values in a column add to a fixed total, such as 1, 10, or 100.

If one of the columns represents a less-the-better factor, the team must convert the factor to more-the-better. For example, if "degree of difficulty" is the factor, it should be converted to "ease." If the less-the-better factor is cost, and the costs can be estimated, then the factor could be the inverse of the estimated cost.

The prioritization matrix, of course, can be used wherever a team needs to rank-order a set of comparable elements, and the team has primarily its own judgments to serve as the basis for rank-ordering.

Analytical Hierarchy Process

The Analytical Hierarchy Process (AHP) is an alternative method for deciding on the key customers, given a list of all possible customers. AHP is a highly developed mathematical system for priority setting. It is explained in detail in Thomas L. Saaty's two books on the subject.[4] We'll only describe a simple aspect of AHP here, for dealing with this limited problem of deciding on the key customers.

The team creates a matrix with the list of all customers along the left side and also along the top. See the example, Diagram 15-2.

The team then estimates the importance of customers pairwise. They enter the degree of importance of Customer Group A compared to Customer Group B in the cell in row 1, column 2. The value they will put in this cell is one of the following:

9   Customer Group A is extremely more important than Customer
        Group B
7   Customer Group A is very strongly more important than Customer
        Group B
5   Customer Group A is strongly more important than Customer
        Group B

3   Customer Group A is moderately more important than Customer
    Group B
1   Customer Group A is of equal importance with Customer Group B

Use reciprocals $\frac{1}{3}, \frac{1}{5}, \frac{1}{7}, \frac{1}{9}$ when the relationships are reversed.

Having made a judgment about the importance of Customer Group A compared to Customer Group B, the judgment of importance of Customer Group
B compared to Customer Group A is automatic; it is the inverse of the first
value, and the group places the inverse in the cell in row 2, column 1. Thus as
the team fills in the cells above and to the right of the cells on the diagonal, all
the rest of the cells are automatically determined.

The team then normalizes the values in each column. To normalize a value in
a column:

1. Compute the total of the values in the column.
2. Replace each value by the result of dividing that value by the column
   total. The result will be a "normalized" set of values that total to 1.

Importance Judgments

	Customer Group A	Customer Group B	Customer Group C	Customer Group D
Customer Group A	1	3	5	7
Customer Group B	$\frac{1}{3}$	1	3	5
Customer Group C	$\frac{1}{5}$	$\frac{1}{3}$	1	1
Customer Group D	$\frac{1}{7}$	$\frac{1}{5}$	1	1
Totals	1.68	4.53	10.00	14.00

Normalized Importance Judgments

	Customer Group A	Customer Group B	Customer Group C	Customer Group D	Raw Weights	Normalized Raw Weights, Converted to Percentages
Customer Group A	0.60	0.66	0.50	0.50	0.56	56
Customer Group B	0.20	0.22	0.30	0.36	0.27	27
Customer Group C	0.12	0.07	0.10	0.07	0.09	9
Customer Group D	0.09	0.04	0.10	0.07	0.08	8

Diagram 15-2. Analytical Hierarchy Process

This method is described in more detail in Section 6.7, with regard to Normalized Raw Weights. The normalized values bear the same proportions to each other as did the original values.

The team then averages the weights in each row and places the averages in the Raw Weights column. Finally, the team normalizes the Raw Weights by dividing the Raw Weight in each row by the Total Raw Weight, and converting it to a percent (by multiplying the normalized value by 100). The Normalized Raw Weights give a good measure of the team's judgment of the relative importance of all the Customer Groups.

Comparison of Prioritization Matrix Method and the Analytical Hierarchy Process

The prioritization matrix method and the Analytical Hierarchy Process each have advantages and disadvantages. The prioritization matrix method generally requires fewer team judgments than does the Analytical Hierarchy Process. For n choices (company groups) and two factors, the number of team judgments required is $2 \cdot n$. For n choices (company groups) and three factors, the number of team judgments required is $3 \cdot n$.

With Analytical Hierarchy Process matrices, for n choices (company groups), the number of team judgments required is $\frac{1}{2}[n \cdot (n-1)]$. As the number of choices increases, the number of required judgments becomes very large. Diagram 15-3 shows these relationships.

Very often, however, Analytical Hierarchy Process judgments are easier and quicker to make than the Prioritization Matrix judgments. This is because Analytical Hierarchy Process judgments are more intuitive, less focused on a single criterion (as described in this chapter—more elaborate applications of AHP provide a method to include any number of criteria in the judgment-making process), and less constrained than Prioritization Matrix judgments. For that very reason, the judgments may also be less consistent than Prioritization Matrix judgments. Some people prefer the types of summary judgments required by the Analytical Hierarchy Process. Others prefer the more structured approach of the Prioritization matrix method. There is no "best" method.

15.3.4 When There Is More Than One Key Customer Group

Often a team will decide there are several key customer groups, and none of them can be ignored. This might be the case if the product is sold to a distributor, who then sells it to retailers, who finally sell it to end users. Many consumer products fall into this category. For example, packaged food, toys, office supplies, consumer electronics, and household hardware are all items that manufacturers normally sell to distributors, who in turn sell to retail stores, who in turn sell to end users.

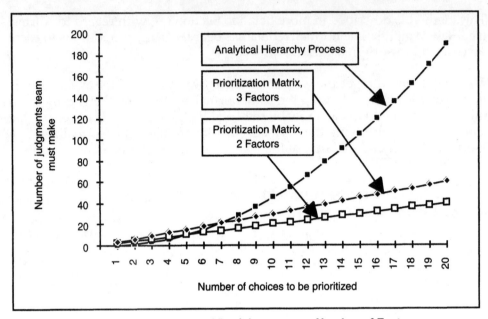

Diagram 15-3. Number of Decisions versus Number of Factors

Customer satisfaction may mean very different things to each of these groups, as indicated in Diagram 15-4.

The success of the product depends on customer satisfaction of the distributor, the retailer, and the end user, and the product must be designed to meet the different needs represented by these three groups.

Customer Group	Needs/Benefits
Distributor	Ease of ordering Just-in-time shipment from manufacturer Easy handling of shipped materials Easy stacking and storing
Retailer	Attractive package Minimum shelf space Product description on package (reduce customer queries) Long shelf-life
End User	Easy to learn how to use Long shelf-life Long product life Safe to use Environmentally correct packaging

Diagram 15-4. Comparison of Different Customer Groups

In this case, some teams have decided to construct a House of Quality that acknowledges multiple customer groups. The general form of the House of the Quality looks like Diagram 15-5.

The product features and capabilities are evaluated against the needs of all customer groups. Besides the obvious increase in the size of the House of Quality, there are complications when dealing with multiple customer groups, especially with respect to the relative weighting of the needs and benefits. We'll have a more complete discussion of these complications and alternative strategies for dealing with them later.

Each customer group has its own set of needs or benefits. Within a customer group, it is possible to develop a rank ordering of needs, by various survey methods. But it is not possible to determine the relative importance of a need in one customer group compared to a need in another customer group.

The reason it is not possible is because we cannot expect a representative from one group to fairly compare the importance of his or her needs to the needs of a representative from another group—these representatives can only speak for themselves.

Instead, the product development team must decide how important it is to them to satisfy each of these groups. Factors that would influence these judgments would be:

- Which customer group's satisfaction is directly related to the greatest short-term revenue or profit potential?

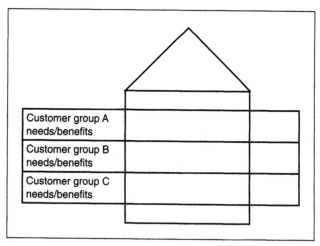

Diagram 15-5. House of Quality for Three Customer Groups

- Which customer group's satisfaction is directly related to the greatest long-term revenue or profit potential?
- Which customer group has the greatest influence on sales of the product? For example, in large companies, decisions on the purchase of office furniture are often made by facilities managers and top management.
- How does communication between the customer groups influence purchase decisions? For example, astute purchasing agents may check with end users on the past performance of a product before authorizing additional purchases of a product or service.
- Does long-term satisfaction of a customer group eventually influence purchase decisions, even if not in the short term? For example, many complaints about the quality or flavor of a cafeteria food item might cause a cafeteria manager to stop ordering that food item.

Any method for deciding on the relative importance of the customer groups would start with the development of a set of criteria that would relate to the groups' importance. These criteria would normally be similar to the list of factors just cited, but could include many other business factors, including business risk, available capital, or ease of accomplishing certain goals.

The goal in deciding relative importance of the customer groups is to develop weights for each group that express the team's judgment of the importance of satisfying each group. These weights can then be multiplied by the importances of the customer needs within each group, to establish relative importances of customer attributes across groups. To illustrate this, look at Diagram 15-6.

Notice that the importances of needs within a group are arranged in descending order of importance *within the customer group,* and that the weights for these needs are normalized; that is, within each customer group, the importances add to one.

In the case illustrated in Diagram 15-6, the team decided to use the ratios 3:2:1 to express the relative importances of the customer groups for their project. These values were determined by one team for a specific application, and they are not necessarily typical of other market segment weightings. By multiplying "Importance of customer group" by "Importance of need within group," the team effectively scaled the importances within a group, so that the resulting weighted importances of need are also normalized. More importantly, the importances could now be compared across groups, and could be resorted in descending order of importance (Diagram 15-7).

The Distributor group was assigned the highest weight, so it is not surprising that most of the needs at the top of the list come from that group. There is one

Customer group	Importance of customer group	Need	Importance of need within group	Weighted importance of need
Distributor	0.50	Ease of ordering	0.32	0.16
	0.50	Just-in-time shipment from manufacturer	0.28	0.14
	0.50	Easy handling of shipped materials	0.24	0.12
	0.50	Easy stacking and storing	0.16	0.08
Retailer	0.33	Attractive package	0.30	0.10
	0.33	Minimum shelf space	0.25	0.08
	0.33	Product description on package	0.25	0.08
	0.33	Long shelf-life	0.20	0.07
End user	0.17	Easy to learn how to use	0.28	0.05
	0.17	Long shelf-life	0.24	0.04
	0.17	Long product life	0.17	0.03
	0.17	Safe to use	0.16	0.03
	0.17	Environmentally correct packaging	0.16	0.03

Diagram 15-6. Weighted Customer Importances

important time-saving observation we can make: one need, "Long shelf-life," appeared in the lists for two groups (End User group and Retailer group). Assuming that these identically worded needs actually are the same need, we can combine them by adding their weights.

The benefit of combining like needs from different segments is a shorter list of customer needs to deal with throughout the QFD process. Reducing the number of rows by even one is a time-saver, because the Relationships matrix is usually much wider than it is high. Therefore, small reductions in row count reduce the overall matrix by quite a bit.

If we choose not to combine identical needs from different segments, we can still save some time in our QFD process by taking note of the these identical needs. Suppose we have two identical needs N_A and N_B, one in segment A, the other in segment B. When we determine the impact of a substitute quality characteristic on N_A, we can automatically use the same impact value with respect to N_B. Because the impact of any SQC on N_A will be the same as its impact on N_B, N_B behaves as a kind of "shadow" customer need, always mimicking the behavior of N_A.

Customer group	Importance of customer group	Need	Importance of need within group	Weighted importance of need
Distributor	0.50	Ease of ordering	0.32	0.16
Distributor	0.50	Just-in-time shipment from manufacturer	0.28	0.14
Distributor	0.50	Easy handling of shipped materials	0.24	0.12
Retailer	0.33	Attractive package	0.30	0.10
Retailer	0.33	Minimize shelf space	0.25	0.08
Retailer	0.33	Product description on package	0.25	0.08
Distributor	0.50	Easy stacking and storing	0.16	0.08
Retailer	0.33	Long shelf-life	0.20	0.07
End user	0.17	Easy to learn how to use	0.28	0.05
End user	0.17	Long shelf-life	0.24	0.04
End user	0.17	Long product life	0.17	0.03
End user	0.17	Safe to use	0.16	0.03
End user	0.17	Environmentally correct packaging	0.16	0.03

Diagram 15-7. Weighted and Regrouped Customer Importances

To complete this example, if we combine the two occurrences of "Long shelf-life," we get a combined weight of .11, and the new overall rank ordered needs become as in Diagram 15-8.

Note that "Safe to use" started with a relatively low importance in the End User group, and after scaling Importance within Group by Importance of Customer Group, it has a low importance overall. Should the developers therefore ignore this customer need? Obviously not. This need is an example of an "Expected Need" in the Klein Grid (Chapter 16). Likewise, safety features are Dissatisfiers in the Kano Model (Chapter 2). Customers generally take safety for granted until a product or service proves to be unsafe, at which point safety will be ranked as extremely important.

15.4　Decide on the Time Horizon

A time frame for shipping a product is a useful constraint for planning purposes. Without a time constraint, a team may set impractical goals for itself. It may decide on objectives that take so much time to complete that the delay in shipping the product outweighs its positive impact on customer satisfaction performance and thereby renders the product noncompetitive.

Thus, a clearly defined time horizon for the QFD process helps keep the planning realistic. The explicit time horizon contributes to better communication by helping all team members to focus on the same objectives.

While it's useful to avoid extremely impractical choices when planning a product, team members can sometimes fall into traps by too quickly ruling out seemingly expensive and time-consuming objectives. Some objectives appear too expensive to be considered at the early stages, but closer study may reveal an alternative, inexpensive approach.

It may be useful during early stages of product planning to regard desirable yet expensive objectives as "breakthrough opportunities." By labeling these objectives as "breakthrough opportunities," the team frees itself up to search for innovative solutions, using methods such as the Pugh Concept Selection process (see Chapter 12).

Customer group	Importance of customer group	Need	Importance of need within group	Weighted importance of need
Distributor	0.50	Ease of ordering	0.32	0.16
Distributor	0.50	Just-in-time shipment from manufacturer	0.28	0.14
Distributor	0.50	Easy handling of shipped materials	0.24	0.12
Retailer/ distributor		Long shelf-life		0.11
Retailer	0.33	Attractive package	0.30	0.10
Retailer	0.33	Minimize shelf space	0.25	0.08
Retailer	0.33	Product description on package	0.25	0.08
Distributor	0.50	Easy stacking and storing	0.16	0.08
End user	0.17	Easy to learn how to use	0.28	0.05
End user	0.17	Long product life	0.17	0.03
End user	0.17	Safe to use	0.16	0.03
End user	0.17	Environmentally correct packaging	0.16	0.03

Diagram 15-8. Weighted and Combined Customer Importances

For planning purposes it's always better in QFD to include all objectives, even the somewhat unrealistic ones, as long as possible, so that breakthrough opportunities are not accidentally discarded.

A good rule of thumb for deciding which objectives to include in the QFD is:

> If the possible objective could be developed by the team by using *all available resources,* during the time frame of the project, include the objective.

In any case, the team must decide on a time-horizon for the project (and therefore for the QFD activity) and use it consistently during the QFD activity.

15.5   Decide on the Product Scope

An important QFD principle—and indeed an important principle in most creative design activities—is to postpone detailed design work as long as possible. In this way the focus remains on the objectives, and the range of possible solutions for meeting those objectives stays unconstrained to allow the most creative and efficient solution to emerge. This means the team must keep the technical solution vague as long as possible.

Nevertheless, there is such a thing as being *too* vague. In any given situation, certain customer needs are likely to be irrelevant. For example, if the product being designed is to be a camera, the customer's needs—and related Substitute Quality Characteristics—outside of those related to using a camera do not need to be considered. Thus, as the Voice Of The Customer is being collected, questions on such topics as healthcare needs or financial management needs would be distractions.

On the other hand, questions on the selection and processing of film, on choices for lighting, and on subjects to photograph, all could provide insight to the development team. The customer needs elicited by these questions could help the team develop Substitute Quality Characteristics that point them to an innovative and competitive camera design.

These remarks indicate a need for *setting the scope* of the QFD activity. The scope defines what's *in* and what's *not in* the QFD discussions to follow. The development team needs to decide for itself, or to determine from its management, how much freedom they have for developing their solution. Knowing the scope helps the team to ignore irrelevant data and to include all relevant data and ideas.

15.6   Decide on the Team and Its Relationship to the Organization

In some cases, the QFD process is used to make decisions that affect a very small group. For instance, most consulting teams that I have been a member of (since we learned of QFD in 1986) have used QFD to determine the relationship the consulting team itself should have with its clients. In cases like this, the entire team affected by the QFD results—the consultants themselves—developed the QFD matrices.

In most cases, however, the QFD results are intended to affect the work of an organization much too large to work all together on the House of Quality.

The team that develops the QFD matrix will be making key strategic decisions about how a product or service should look or be. Every good manager knows that the people implementing a decision will be more motivated, better informed, and generally more able to do the job if they had a part in forming the decision in the first place. The general principles that come into play are "ownership" and "informed decision-making," both of which are strongly enhanced in people who move in the direction they themselves have set. Thus, when product or service planning employs QFD, it's best for the implementors to be as closely involved in the initial planning as is practicable.

Beyond that, the ideal QFD team should include representatives from all the important functional groups involved in designing, developing, building, delivering, and servicing the product or service. There's a tendency for product developers to exclude certain functional groups from their initial product planning—a relic of the old "throw it over the wall" method of product development—on the grounds that those whose responsibilities are downstream from the initial work lack the knowledge or expertise needed to usefully influence the upstream decisions. In a sense, however, QFD stands for the opposite point of view.

In the QFD paradigm, everything important about a product or service gets decided at the beginning. Whatever customer satisfaction aspects of the product are ignored upstream are likely to be difficult or impossible to fix downstream. Hence, the QFD team composition is crucial to the overall success of the product or service.

For manufactured products, a typical list of functions that should be represented on the QFD team would look like this:

- Marketing (and in certain settings, representatives from the customer)
- Sales (if separated from Marketing)

- Product design
- Supplier management/Purchasing (and sometimes, representatives from key supplier organizations)
- Manufacturing engineering
- Manufacturing production
- Order processing/fulfillment
- Service

Some of these functions may be themselves multifunctional. For example, for computer design, various subfunctions might include:

- Processor/logic design
- Packaging and power design
- Design tools (CAD) design
- Product documentation

For many marketing functions, the subfunctions might include

- Domestic marketing
- Overseas marketing
- Market research
- Sales support and liaison

Allowing for such multifunctional subdivisions in other functions as well, the list of possible representatives may be far too large. This situation occurs very often in larger organizations.

A high-level manager would probably appoint representatives from each function to represent all the subfunctions in their group. These representatives would then have the responsibility of communicating the QFD teams' work to their constituents. Criteria that would influence the choice of team members would be

- Knowledge of the customer
- Experience in developing similar products or services
- Willingness to participate in disciplined planning process implied by QFD
- Accountability for the subsequent development work
- Authority to make decisions and commitments on behalf of the organization represented
- Enough knowledge to represent the concerns and issues of the organization represented

An alert QFD facilitator will assess the cultural norms for communication in the client's organization. At the outset, he or she will establish mechanisms for information to flow from the QFD team to the key parts of the organization that will be affected by the QFD results.

The mechanisms will also allow for reactions, opinions, and data to flow from the outside organization back to the QFD team. A healthy two-way flow of information between the QFD team and the rest of the organization assures that the QFD results will not come as a surprise and will not create undue controversy. This, in turn, assures that the QFD team's results will be rapidly accepted and more easily implemented by the organization.

One team I worked with was using QFD to develop a set of top-level product development criteria for several families of new products to be developed in their rather large corporation. The QFD process was scheduled for five one-day and two-day meetings over a period of six weeks.

The QFD team's results were expected to create some radically new ways of thinking about the company's products. The goal was that these results should cause a minimum of controversy, and take no one by surprise.

At the outset, each team member took the responsibility for briefing a key manager or Vice President in the company after each one-day or two-day QFD meeting. The team members were to explain the major issues the QFD team was working on, and what new or surprising ideas or points of view the QFD team had developed. They were also expected to get the manager's reaction and bring that back to the QFD team. In this way, a two-way dialogue occurred between the QFD team and the rest of the corporation, as the QFD team did its work.

As the QFD effort continued, the team did indeed develop many novel ways of expressing product performance (based on new customer data). These novel ways were bound to entail new measurement methods and new approaches to benchmarking the competition. Rather than inciting controversy, however, the communication process set up by the team had the effect of creating *excitement* around the company. People who were unable to work on the team were impatient for the results, and in one case began making their own product plans based on partial results of the QFD team, who had not yet completed their work.

15.7   Create a Schedule for the QFD

The first time I ever explained QFD to a product development team, I was asked how long the QFD activity would take. I hadn't the foggiest notion, having never been through the process myself. I learned the hard way, working with that team and others shortly thereafter, that

- QFD requires time
- QFD can be shortcutted
- QFD should be a managed activity, just like any project

15.7.1   QFD Requires Time

Every one of the steps detailed below could be a part of a QFD process. Every one of them takes time to perform. The most rapid product QFDs I have facilitated took about two days. Drastic surgery was done to the QFD process in order to keep the process that short, and it was done only because I judged that within the environment of this team's operations, the QFD team could not allocate more time. It seemed to me that two days of QFD-style product planning was better than no structured planning at all.

In the QFD Estimator Chart (Diagram 15-9), I've identified the most common steps in a QFD process for developing the House of Quality. Next to each step I've provided some rules of thumb for estimating the length of the step. The third column in the table shows typical sizes of tables or data for the step, and the fourth column shows the resulting estimated time for the step. The rest of this section provides explanations for the contents of this diagram.

The steps are described in much more detail throughout this book; I've added comments here to explain the rules of thumb and how they lead to the estimates I've provided.

QFD Estimator Chart

Some general rules of thumb seem to hold for several of the important steps in the QFD process. Some of these have to do with the amount of data to process, others with the rate at which teams can process the data. A good way to estimate the time needed for QFD can be based on the following rules of thumb:

1. For any market segment, the number of tertiary customer needs will usually be between 75 and 150.

2. For any market segment, the number of secondary customer needs will usually be twenty to thirty.

3. Most teams will generate about three Substitute Quality Characteristics for each secondary customer need.

4. Smaller teams will process data much faster than larger teams. The fastest processing can be done by a single person—no arguments at all are likely to occur! The larger the team, the more likely it is that there will be differing points of view, and therefore the more likely it is that there will be discussion. One of the important benefits of QFD is the common understanding the team develops of the customer and of the problem they are trying to solve. This common understanding is the natural by-product of the disagreements and the resultant discussions. However, the development team will have to balance the benefits of the discussions against the time they take. Larger groups mean more discussions and deeper common understanding. Smaller groups mean faster QFDs.

Activity	Rule of thumb	Typical size	Typical time
Decide on the customer	$\frac{1}{2}$ day discussion	15 customer categories identified, 3 are key categories	$\frac{1}{2}$ day
Gather the qualitative needs	1 hr. per interview, 4 interview locations, 3 hrs. analysis for each hr. of interview	20–30 interviews	15 days
Structure the needs	Team does it: $\frac{1}{2}$ day. 200 customers do it: 3 weeks.	150 unique needs at tertiary level, 25 needs at secondary level	$\frac{1}{2}$ day to 3 weeks.
Quantify the needs	Team does it: $\frac{1}{2}$ day. Customers do it: 1 to 3 months.	25 needs at secondary level	$\frac{1}{2}$ day to 3 months.
Set performance goals	25 secondaries, $\frac{1}{2}$ day	25 secondaries	$\frac{1}{2}$ day
Generate Substitute Quality Characteristics	3 SQCs for each secondary attribute: 2 days for 25 secondary attributes	25 secondaries, 75 SQCs	2 days
Determine impacts—SQCs to needs	1 minute per SQC/ secondary pair	25 secondaries times 75 SQCs = 1875 cells	All cells done by entire team: 3.9 days Reduced matrix assigned to teams: 2 days
Determine technical correlations— SQC to SQC	1 minute per SQC/ SQC pair	$\frac{75 \cdot 74}{2} = 2775$ comparisons	All cells done by entire team: 5.8 days Reduced matrix assigned to teams: 2 days
Benchmark	Varies widely, depending on products, services, technology, and market conditions		
Set targets	Varies widely, depending on products, services, technology, and market conditions		

Diagram 15-9. QFD Estimator Chart

5. Teams always work very slowly at the beginning of the QFD process. Contributing factors are lack of familiarity with the process and lack of experience with the thought processes that the QFD discipline promotes. In many cases the QFD team members may not know each other well. Even if they do, they may be unaccustomed to interacting with each other on the range of topics and type of customer needs that QFD deals with.

The time estimates provided below represent rough averages of time periods that actually vary widely, starting with very long time periods (half a day to make a single decision) when the team is inexperienced, and ending with very short time periods (two or three minutes to make a single decision) as the team gains experience and confidence.

6. Teams almost always work very rapidly as they near the end of the process. Contributing factors are increasing familiarity with the QFD process; increased familiarity, comfort, and trust among the QFD team members; and development of a vocabulary and concept kit that simplifies QFD judgments.

Here are the main steps in the QFD process. Special considerations affecting time schedules are detailed with each step. The QFD Estimator Chart (Diagram 15-9) summarizes the major points involved in scheduling.

Customer Needs / Benefits Matrix

- **Decide on the customer.** There is usually quite a bit of disagreement at first in deciding on the key customers. In my experience, teams have spent anywhere from one to eight hours discussing this point. In some cases, the question is not settled by discussion, but rather by sizable market research studies aimed at defining useful market segments. Such studies can take weeks or months. The QFD Estimator Chart (Diagram 15-9) assumes the issue can be settled by discussion.

- **Gather the qualitative wants and needs.** There are so many methods for doing this that no single guideline can apply. Assuming the method involves interviewing, I suggest you estimate the length of the interviews, take travel into account, and provide time for analysis of the interview data. The primary output of the analysis would be the customer wants and needs (at the verbatim [tertiary] level). There is usually much more information available from these interviews than just the wants and needs. For estimating purposes, however, I've excluded analysis of everything but the wants and needs.

In the table, I've assumed one hour per interview, twenty-five interviews, four days of travel (amounting to forty-eight person-hours), and three person-hours analysis for each interview hour.

- **Structure the needs.** If the development team creates the structure by using the affinity diagram process, 100 to 300 needs can usually be structured in two to six hours. I have worked with groups that started with as many as 750 needs: the process took about eight hours, and it was exhausting. As with all group processes, the smaller the group, the faster the process. You can include customers in the group and still get the task done within these time frames.

  The best method for structuring needs—having many customers do it, and then constructing a composite affinity diagram representative of all the customers—is generally much more time-consuming, although more representative of the customer, than any other way to structure the needs.[5] For more on methods for customers to affinitize customer needs, see Section 16.5.1. Factors that affect the schedule are

  > Time to recruit customers
  > Number of customers participating in the process
  > Mailing mechanism for sending the needs to be sorted to the
  >   customers and receiving them back
  > Method used for analyzing the received sorts

Given the right statistical tools, this process can be done in as few as three weeks, but more commonly about ten weeks.

Planning Matrix

- **Quantify the needs.** At the secondary level in the hierarchy, determine the quantitative importance of the wants and needs to customers, the company's performance relative to meeting those needs, and the competition's performance. The easiest way to do this is to "invent" the numbers—that is, to use the judgment of the QFD team to determine the importance and performance numbers.

  This process usually involves considerable debate within the QFD team. When a team is making judgments, there can be no "right" answer other than consensus. Therefore, the goal is consensus. Most teams can come to consensus on importance, satisfaction performance, and satisfaction performance of the competition for twenty to thirty secondaries in about half a day. With a strong facilitator, most teams can come to consensus within any time frame they choose.

  The other approach to quantifying the needs is to collect data from the customers, usually by survey, as discussed in Chapter 6. This process assumes that the qualitative research has already been completed—that is, the customer wants and needs have already been determined. Given a set of wants and needs, customers are asked to rank-order them, or to identify the few most important needs. They are also asked how well their current vendor met each need. If a sufficient sample is chosen, enough customers will respond to fairly represent satisfaction performance with the competition.

Parameters affecting the time required to survey customers are

Number of wants and needs
Number of customers surveyed

The number of wants and needs customers are asked about must be kept fairly small (conventional wisdom suggests twenty to thirty). Larger numbers require more time than survey respondents are willing to give. Larger numbers are also confusing to customers; they often find they cannot easily differentiate among as many as forty or fifty needs. This is one of the reasons that QFD is usually done using the customer needs at the secondary level: The number of secondaries rarely rises above twenty-five to thirty for a single market segment.

The number of customer needs influences the time required to develop the survey, the time customers need to complete the survey (and therefore the return rate), and the time to perform analysis of the survey results.

The number of customers surveyed, often called the sample size, is critical to the confidence people will have in the results. In general, the more customers surveyed, the more survey results will be returned or responded to. However, the QFD team will normally balance the cost and the time required to survey larger numbers of customers against the advantages of higher confidence levels.

The following factors are affected by the sample size:

Time to recruit survey respondents
Time to present the survey to customers (time to print and mail survey
    forms, or to make telephone calls, or time to visit customers)
Time to analyze the survey results

· **Set satisfaction performance goals.** This group consensus process has a far-reaching impact on all the QFD results. As with other consensus processes, there is no right answer. As a consensus process, the factors that affect time to complete are the size of the QFD team, the abilities of the group facilitator and the number of items to come to consensus on. It is rare to spend more than half a day on this process; frequently only a couple of hours are required.

Substitute Quality Characteristics

· **Generate Substitute Quality Characteristics (SQCs).** This process is quite variable. It depends on many factors:

The nature of the SQCs: performance measurements or metrics; product requirements; or product features of capabilities

How SQCs are generated: from the customer wants and needs; or already available (from prewritten Product Requirements Documents, or from a previous QFD process, for example)

In general, if the QFD team must create ideas during the QFD process, it takes longer than if the ideas had already been prepared. Whether the ideas are created *before* QFD or *during* QFD, it still takes time to do the work. However, we're accounting for the QFD time itself, so we've made that distinction. The QFD Estimator Chart assumes the team will generate SQCs during the QFD (this is the most common practice).

When the wants and needs are used as the basis for generating SQCs, the clarity and purity of the SQCs is very important. Assuming they are clear to the QFD team, and each one expresses single, not multiple needs, my rule of thumb is that the team generates an average of 3 performance metrics for each customer want or need. The work often starts extremely slowly (the first performance measure might take 1/2 day), and then accelerates as the team's confidence grows.

Impacts

- **Determine impacts of Substitute Quality Characteristics upon needs.** Once the team understands what type of judgments they must make for this step, the process generally moves quickly. The requirements for a quick process are, of course, that the customer wants and needs are clear and unambiguous to the team, and that the SQCs are also clear.

Often the relationship matrix is sparse—that is, the number of high and medium impacts is small compared to the total number of cells. When the matrix is sparse, the work goes rapidly, because the team can usually agree in a flash when no relationship exists.

Two important time-saving strategies are available to the team:

dividing into subteams

filling in a reduced-size matrix

Both strategies are described more fully in Section 15.7.2.

When the team divides into n subteams, the work takes much less than $1/n$th the time of the whole team working on the matrix together. The new element that accounts for this greater-than-expected increase is that smaller teams are able to come to consensus much more rapidly than large teams.

The reduced-size matrix—the matrix consisting of only those rows relating to the most important customer wants and needs—is often one-third the size of the full matrix, and therefore requires proportionately less time to complete.

Technical Correlations

- Assess the strengths of supporting or opposing correlations among Substitute Quality Characteristics. In terms of time estimates, this process is similar to the process of determining impacts.

  Quite of few of the correlations cells will be empty and will be rapidly processed by the team. Some teams limit their analysis only to the highest priority SQCs. Some teams assign the work to subteams. Many teams skip this step entirely. Some teams limit the analysis to the most highly ranked SQCs.

Perform Competitive Benchmarks and Set Targets

- Determine, either by benchmarking the competition, by laboratory experiment, or by some other method, what targets to set for the key Substitute Quality Characteristics. The nature of this work varies so widely that no reasonable time estimate can be given.

15.7.2   QFD Can Be Shortcutted

In Section 15.7.1, "QFD Requires Time," we mentioned two techniques for shortening the process. They are

- Matrix reduction
- Dividing into subteams

Let's look more closely at these techniques.

Matrix Reduction

Many QFD practitioners advocate the use of small matrices in QFD. They suggest that rather than spend all available time filling out a single large House of Quality, the team would gain more benefit by filling out just the critical parts of the House of Quality, and then just the critical parts of other matrices for downstream parts of the development process (such as the Design Deployment and Manufacturing Planning Matrices, described in Chapter 18).

In order to keep the HOQ small, some choices must be made to eliminate some of the detail in the Voice of the Customer. By working with fewer rows of "Whats," the team will generate fewer columns of "Hows." Even if the team has already generated a lot of "How" columns, just reducing the number of "What" rows will substantially reduce the matrix size. Besides selecting a higher level of the VOC hierarchy to start with (as described in Section 3.2), there are two other ways of reducing the number of Voice of the Customer rows: judgment and raw weight analysis.

Using Judgment to Reduce Matrix Size

The development team may be working on a focused problem that is more narrow than design of a total product. In this case, they can decide which customer needs pertain to their problem, and drop the rest. The team could develop a prioritization matrix (Section 3.5), where a significant criterion used for rank-ordering customer needs would be "Applies to our problem."

In one popular method, sometimes called the Nominative Group Technique (a term that applies generically to many group voting processes) each member of the team is given a certain number of votes, perhaps five votes. They can cast their votes for as many or as few Customer Needs as they like, including casting all of their votes for a single Customer Need. The highest scoring customer needs will then be the ones ranked most important by most of the team members.

Another similar method is called the Multipickup Method (MPM.) It has been described by Professor Shoji Shiba.[6] In this method, a selection round involves each team member marking those Customer Needs they consider worth retaining. At the end of the round, all unmarked Customer Needs are removed.

Using Relative Raw Weights to Reduce Matrix Size

In any prioritization matrix (of which the Relationship matrix is one example), the more heavily weighted "Whats" will have a greater influence on the weighting of the "Hows" than will the less heavily weighted "Whats." We can use that simple fact to cut significant work out of the QFD process. Consider the prioritization matrix in Diagram 15-10.

The rank-ordering of the "Whats" and "Hows" is shown in Diagram 15-11a.

Suppose we ignore the impacts in the matrix row containing the least important "What" (Row F)? How would that change the ordering of the "Hows"? By setting the importance of the least important "What" to zero—which is equivalent to not filling in the impacts in that row—we can see (in Diagram 15-11b) that the rank ordering and the weights of the "Hows" has changed very little. In fact, it may take you a moment or two to see any change at all.

If we ignore the *two* least important "Whats," D and F, the change in rank order and the weights of the "Hows" is more noticeable (Diagram 15-11c), but the three most important "Hows" are still at the top of the list, while the three least important remain at the bottom of the list.

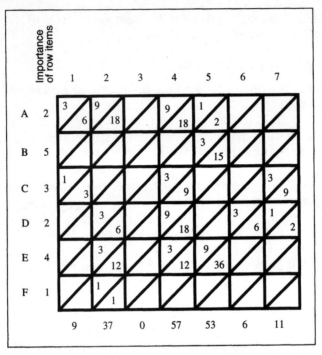

Diagram 15-10. Prioritization Matrix

Taking the process a step further (Diagram 15-11d), and setting the importances of the *three* least important "Whats" to zero, we continue to see relatively little shift in the rank ordering, although the relationships between normalized importances have indeed changed quite a bit.

The more of the less important "Whats" we drop out, the more the resulting importances of the "Hows" diverge from the original case, where all the "Whats" are taken into consideration. However, the divergence is gradual. By dropping a single row, sometimes very little of the final result, the rank ordering of "Hows," changes. To the extent that our goal is just to prioritize the "Hows," we can sacrifice some detail in the rank ordering of the "Hows" and still get the prioritization that relates strongly to high impact on the "Whats."

All this leads us to this time-saving strategy. If there may not be enough time to complete the matrix, be sure to begin evaluating the highest ranking "Whats" first. As you get closer to completion, the rank ordering of the "Hows" will not change very much and you can safely drop off some of them.

To get an idea of the most dominant "Whats," arrange them in descending order of importance and compute the normalized importance and cumulative normalized importance.

What	Importance	How	Importance	Normalized importance
B	5	4	57	.33
E	4	5	53	.31
C	3	2	37	.21
A	2	7	11	.06
D	2	1	9	.05
F	1	6	6	.03
		3	0	.00

Diagram 15-11a. Relationships Using All "Whats"

The "Whats" with Cumulative Normalized Importance (B and E in Diagram 15-12) greater than approximately .50 dominate the entire set of "Whats" and should always be included in the matrix. Beyond those, the more "Whats" the better. If you want to estimate the effect of the lower 50 percent of "Whats," you can generate a worst-case scenario in which you set impacts for all "Whats" in the lower 50 percent to the highest value (usually 9). Compare the resulting normalized weights of the "Hows" to the case where the impacts are set to zero. If there's a big shift in normalized importances, you must complete more of the HOQ. If there's very little shift, you can safely conclude that you will learn very little new by completing more of the HOQ. Be sure to use *normalized* weights when looking at the importances of the "Hows." Absolute values are guaranteed to change, but the relative differences between the weights are best represented by normalized values.

Typically, for one customer segment, a HOQ will have about 25 "Whats," and the normalized importance of about the top one-third of them will accumulate

What	Importance	How	Importance	Normalized importance
B	5	4	57	.33
E	4	5	53	.31
C	3	2	36	.21
A	2	7	11	.06
D	2	1	9	.05
F	0	6	6	.03
		3	0	.00

Diagram 15-11b. Relationships Eliminating Least Important "What"

What	Importance	How	Importance	Normalized importance
B	5	5	53	.38
E	4	4	39	.28
C	3	2	30	.21
A	2	7	9	.06
D	0	1	9	.06
F	0	6	0	.00
		3	0	.00

Diagram 15-11c. Relationships Eliminating Two Least Important "Whats"

to 50 percent of the total normalized importances. Most HOQs are considerably wider than they are tall, so removing a few "Whats" from the matrix usually amounts to substantial reduction in matrix size. In Diagram 15-13, by eliminating the three least important "Whats," we have reduced the Relationship matrix by $3 \cdot 17 = 51$ cells, with very little loss of critical information.

Dividing into Subteams

It is a common adage in organizational dynamics that in order to reach consensus, a group of n people must have n^2 conversations—that is, each person must have a conversation with each other person in the group. While most groups are not too careful to compute and keep track of the number of conversations or interactions they have, it is certainly true that the larger the group, the more time it takes for the group to come to consensus on any issue. (Of course, there is no guarantee that a particular group *will* come to consensus on a particular subject!)

What	Importance	How	Importance	Normalized importance
B	5	5	51	.53
E	4	4	21	.22
C	3	2	12	.13
A	0	7	9	.09
D	0	1	3	.03
F	0	6	0	.00
		3	0	.00

Diagram 15-11d. Relationships Eliminating Three Least Important "Whats"

What	Importance	Normalized importance	Cumulative normalized importance
B	5	0.29	0.29
E	4	0.24	0.53
C	3	0.18	0.71
A	2	0.12	0.82
D	2	0.12	0.94
F	1	0.06	1.00

Diagram 15-12. Normalized Importance of the "Whats"

Taken at one extreme, consider the group of size one. Most individuals can make judgments for themselves—at least the focused individual judgments that come up in QFD—very rapidly. No discussion is needed. None of the arguments in favor of a particular opinion need be debated; they can be reviewed within the mind of the single group member almost instantly.

With a group of size two, the situation is quite different. All the private thoughts that a single person did not have to discuss with himself, or herself, now must be articulated to the other group member. The cost in time to

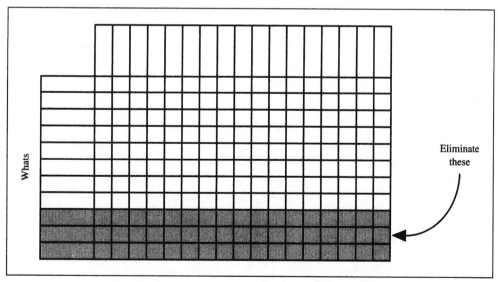

Diagram 15-13. Eliminating Least Important "Whats"
("Whats" in Descending Order of Importance)

communicate these thoughts is extremely high compared to a single person "talking to himself," and the chances for misunderstanding are much higher. Misunderstanding in turn leads to the need to explain the ideas yet again, thus requiring even more time. If these costs were not already high, let us not forget that the communication must go in both directions. Each person must communicate his ideas to the other, and both together must converge to a consensus.

With groups of three or more, the number of interactions certainly goes up, even if not as the square of the size of the group. Furthermore, the larger the group, the less intimate the environment, so that the quality of the communication must become more impersonal and more formal. Finally, each person gets less time to state a position, further contributing to misunderstanding.

For all these reasons, the way to get decisions made quickly is to assign them to individuals. If speed were the only requirement when doing QFD, we would ask a single person to fill in the House of Quality—it could probably done in just a few hours.

The largest matrix (in terms of cell count) in the House of Quality is the Relationships section. On my first QFD project I failed to perform the obvious simple operation of multiplying the 110 customer needs by the 155 Substitute Quality Characteristics we had generated. Had I done so I would have been profoundly impressed by the resultant product (17,050). It is almost impossible to get a team to sit together in a room and come to agreement on 17,050 judgments. Even if the team had had the patience and fortitude to come to consensus 17,050 times, the time to complete the task would have been overwhelming and unacceptable to them. Suppose that an average of three minutes would have been needed for discussing each cell. The process would have taken 852 hours, or 106 full working days.

Diagram 15-14 illustrates the impact of using subteams on the overall time required to fill in the Relationships section. Suppose that I had asked each of the ten people on the team to fill in a portion of the Relationships section on his own. Most people can, on their own, make the required judgments at least as fast as one per minute. As one becomes familiar with the elements on the left and on top of the matrix, one can make the judgments much faster, usually three or four per minute. Assuming a conservative rate of one per minute, with ten people working in parallel, the team could complete the 17,050 cells in only three and one-half days.

Suppose instead that the group had been divided into five groups of two people. The average time to make each judgment might be twice as long as an individual would need—say two minutes. The overall time to make the 17,050 judgments jumps to fourteen days.

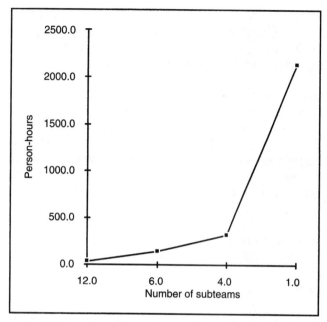

Diagram 15-14. Person-hours to Complete 17,050 Cells with 12 People, Divided into 12, 6, and 4 Subteams, and Working as a Single Team

A matrix of 17,050 cells is an extreme example. I have never seen a matrix that large be completed. But these simple calculations suggest how much the overall time is affected by dividing the QFD team into subteams.

Of course, completing QFD rapidly is *not* the only requirement. (If it were, the quickest way to complete QFD is to not do it at all!) The quality of the QFD decisions is an extremely important requirement, as is group buy-in or development of a common group vision. Since the surest path to group buy-in is to give the group plenty of opportunity to discuss the problems, the QFD facilitator must continually balance the benefits of group discussions against the cost in terms of time.

If we do divide into subteams, how can we optimize the quality of the decisions? Most groups adopt a hybrid strategy: perform the task partly as a single large team, and partly as subteams. When I facilitate QFDs, and decide it's necessary to divide into subteams, I usually have the entire team work together for a few hours, to help create a "standard" for how teams will determine impacts.

As facilitator, I can help the group get accustomed to the questions they must ask and answer in order to make the judgments that QFD calls for. If there is a problem in the matrix—for example, unclear Customer Wants/Needs, or imprecise Substitute Quality Characteristics—there's a good chance the problem can be dealt with while the entire group is assembled. That will make for smoother sailing when the group breaks into subteams.

If possible, I have the subteams work together in different parts of the same room. Then, if one team runs into a problem or discovers the need for clarification or addition of a Substitute Quality Characteristic, it can be easily and rapidly communicated to the other subteams. I have also found it helpful for each subteam to present a few of their judgments to the other subteams at "checkpoints" along the process. This ensures that each subteam is making judgments that the other subteams have confidence in.

In the discussion above, I have focused on the Relationships matrix as the place to break into subteams. However, it's not the only place where the QFD time can be effectively reduced. Any part of the HOQ where team judgments must be made is an effective candidate to be sped up by the use of multiple smaller teams working in parallel.

However, it's a good idea to have the entire team work together at the beginning of the process, no matter how slow the progress. The team needs to develop a mutual understanding of the QFD process and of each other before they will have confidence that breaking into smaller groups will produce results acceptable to the whole team.

It's important to bring the subteams back to work as a single team often enough so that the sense of "teamness" remains. Finally, as the HOQ is completed, the team must analyze the results and determine a course of action. This must be done with all team members together.

15.7.3   QFD Should Be a Managed Activity

After facilitating several QFDs, I came to realize that the amount of work in each of these steps can be estimated in advance. I also discovered that the process can use up the time of many highly paid people—the QFD team members. As a facilitator I was responsible for helping them to use their time effectively. The best way for me to do that was to lay out a detailed schedule for the entire QFD process, making use of estimates of the size and number of QFD matrices that were likely to be built during the process.

I eventually adopted the practice of establishing a detailed schedule, sometimes predicting task completion to a specific hour. The schedule helped to create a sense of time pressure among the team members, and effectively discouraged the team from wasting time.

I recommend that any facilitator lay out the QFD process just as if it were a complex project. Based on the objectives of the team leader, determine the number of matrices that will be constructed, and other activities that will be part of the process. Make assumptions about the sizes of various matrices (see Diagram 15-9, the QFD Estimator Chart), whether the team needs to divide into subteams, and the average time for the team or subteam to make each judgment. Lay out the tasks on a GANTT chart or Task-Dependency chart (such as an Arrow Diagram). Estimate start dates and start times.

Don't be afraid to microestimate the process. It's like establishing the agenda of a long meeting, or a series of meetings. You plan at what time events start and end, in which room, and with which individuals. Things may not go as you plan; some things will take longer, others will go more rapidly. But the detailed schedule gives the facilitator a guide for determining when to intervene and reschedule. The overall schedule helps establish the team's expectations, and thereby helps keep people motivated for the entire process.

Most importantly, the carefully completed time estimate is a key to ensuring that the QFD will be completed within the time frame people expected, and will achieve the desired objectives.

15.8   Acquire the Facilities and Materials

15.8.1   Location

The QFD process can be spread out over many days or weeks, or it can be concentrated in just a few days. In either case, the team will find the process absorbing and energy-consuming. Unscheduled interruptions should be avoided, as they seriously slow down the process and lower the quality of the results.

To encourage concentrated involvement, many teams choose to locate the QFD activities away from their normal workplace. They discourage incoming phone calls, and try to make it difficult for team members to make outgoing calls, even during breaks.

Occasionally, however, the team will need to refer to materials located away from the QFD site. These materials might be marketing reports, product test analyses, or material that will help the team answer some question that has come up.

If the team has chosen to locate the QFD activity away from the workplace, it will be helpful to establish some method for accessing materials that normally reside at a distance. Access methods could include facsimile devices, couriers, or a computer connected to some critical project database. Without such an access method, the work of the team can be delayed.

15.8.2 Room

The room chosen for QFD must have plenty of bare wall space. QFD matrices are often three to five feet high, and ten to twenty feet long. Entries are made to the matrix while it's hanging on the wall. Architectural columns, corners, molding, and pictures over which the matrix is hung will only make the matrix difficult to read and to write on.

The wall surfaces must allow for hanging of the QFD matrices with tape or another adhesive, or with pushpins. Some facilitators like large "white boards" because Post-it Notes stick directly to them. This author prefers Post-it Notes placed on large sheets of paper that can be removed, with the Post-it Notes still in place on them.

There should also be comfortable chairs and adequate working surface space for people to lay out papers or sections of the QFD matrix.

The room should be brightly lit, and the acoustics should allow people to hear each other easily. The author facilitated one QFD process in a room with extremely high ceilings. People's voices echoed enough so that voices tended to "get lost" in the room. Often one person was unaware that another was speaking. People interrupted each other without realizing that they were doing so. The meeting frequently broke down into several smaller meetings, occurring simultaneously.

It's best to use the same room for the entire QFD. Assuming the room contents can be kept away from people with no need to know, much time can be saved, and continuity can be preserved, by not having to take down and put up the wall charts every day.

Whenever possible, inspect the QFD room in advance. If it's not ideal, you'll have time to figure out how to make the room work for you, or to find another room.

15.8.3 Computer Aids

The matrices of QFD are repositories for hundreds or even thousands of bits of information, including qualitative data (such as customer wants and needs), quantitative data, narrative text (such as assumptions that explain the data, decisions that the group made, or issues yet to be resolved), and even graphical data (such as diagrams of concepts in Pugh Concept Selection Matrices).

Today's desktop and laptop computers run plenty of excellent software products that can help store, manage, and display this data.

From very early on, a few enterprising software companies developed specialized software to support the QFD process. This software allowed the calculations to be customized, the matrix dimensions to be varied, and the matrix to be printed in a format that looks like a "textbook" QFD matrix. The various software products are more or less easy to learn and use, and they run on computer platforms that meet most people's needs.

Beyond these specialized software products, computer spreadsheets can be customized to store the QFD data, compute the formulas, and print the matrix.

The following are critical advantages of using computer software for QFD:
- It serves as a central repository for all decisions, judgments, comments, issues, and notes generated during QFD.
- Computations are performed automatically, and, one may presume, correctly.
- It greatly simplifies, perhaps even makes possible, "what if" scenarios, such as setting importance of certain wants and needs to zero; setting impacts of certain cells to "High"; or substituting various values for "High/Medium/Low" impacts (9-3-1, 7-3-1, 5-3-1, 3-2-1, for example).
- The matrix can be printed in a relatively compact format (several sheets of $8\frac{1}{2}''$ by $11''$ paper, for example).

A useful way of using QFD software is to assign a notetaker or scribe to operate the computer in the QFD room. This person would be seated out of the way of the group's activities, yet still be able to see the QFD chart hanging on the wall. The scribe would be familiar with the operation of the computer and software, and would quietly record data into the computer as the team makes its decisions. When needed, the scribe can simply read out the results of computations, which the facilitator can copy onto the QFD chart on the wall.

Some people have attempted to use on-line LCD devices that can project the computer screen onto a projection screen in the conference room. I have found this unsatisfactory with today's technology for two reasons:
- Projected LCD images are not very bright—therefore, the room must be darkened somewhat. This creates difficulties in other areas because people cannot effortlessly view both the QFD chart on the wall and their papers.
- Today's computer screens, especially the LCD projection devices, can only display a small fraction of the QFD chart at one time. Team members lose the ability to scan the entire chart at will, looking for patterns or referring to previous decisions and judgments.

Many QFD teams prefer that each team member have a small, personal, up-to-date copy of the QFD chart. The computer can print these charts frequently, and photocopies are usually easy to make. This makes the need for a QFD chart on the wall or projected on the screen slightly less important.

I look forward to the day when computers will be able to display the entire QFD chart on a large wall, in a brightly lit room, at a reasonable cost! Until then, I recommend using the computer to record results, but maintaining the full QFD chart where everyone can inspect any part of the chart they wish, at any time.

15.8.4 Materials

The QFD process is team-oriented. The team works together, developing the information to be filled into the matrix. To facilitate this team process, the matrix must be visible to everyone on the team, all the time. Ideally, each team member should be able to scan the matrix from top to bottom, side to side, whenever he or she wants to.

The most commonly used materials for QFD do a pretty good job in meeting these needs. Fortunately these materials are inexpensive, readily available, and familiar to everyone who works in an office.

Flip chart pads provide large, but still manageable, pieces of paper that can be pieced together to form matrices of any size. Some pads have pre-ruled grid lines printed on them. The grids are usually an inch wide. These preprinted grids obviate the need to manually rule lines, something the author did often, with makeshift straight-edges, while the QFD team was taking a break or driving home for dinner. I strongly recommend paper with preprinted gridlines.

Masking tape or pushpins are needed for hanging the matrix on a wall, so everyone can easily see it. (Obviously, walls that don't allow for hanging the matrix are to be avoided.) Even if the matrix is held up by pushpins, it's often desirable to tape the sheets together at all the edges, to construct a single large chart. This tactic ensures that the sheets won't get out of order or lost if the matrix has to be hung, brought down, and rehung several times, as is often the case.

Thick-nibbed markers of various colors are used for recording information in the matrix so that they can be seen clearly from a distance. The ability of the team to view the entire matrix, with its patterns and interactions, cannot be overemphasized. To allow a reasonable "bird's-eye" view, the entries must be large and dark.

Some teams favor various color coding schemes for representing the customer needs segments, or for distinguishing main categories of SQCs, or for recording impacts.

Post-it Notes are invaluable for many purposes. They facilitate team members recording data in parallel; they allow for information to be moved from one part of the room to another, or from one part of the matrix to another, without being rewritten; and they allow for information to be organized, grouped and regrouped as the team's understanding of the relationships between units information evolves.

Summary

In this chapter we have seen what's involved in getting ready for QFD. Planning for a QFD is the single most important action one can take to ensure a useful, timely outcome. A good job of planning involves anticipating the entire process and ensuring that team members' expectations have been appropriately set. It's probably better to err on the side of too detailed a plan rather than too vague a plan.

A good QFD plan involves getting answers to many questions, divided into categories. We've covered the categories of questions by dividing the chapter into these sections:

- Organizational support
- Objectives of the QFD
- Determination of the customer
- Determination of the time horizon for the project being planned
- Determination of the product or service concept to an appropriate level of detail
- Determination of the QFD team and how it will communicate to the rest of the organization
- Schedule for the QFD activities
- Acquisition of the facilities and materials needed

As we have looked at these general areas of questions, we've also had to touch on many implementation aspects of QFD. In a sense, planning is like living through the process.

Having established our plan, the next important step is to listen to the customer. We first discussed this in Chapters 5 and 6. Now we're ready to actually meet the customer face to face. How shall we proceed? Chapter 16 provides the answers.

Discussion Questions

If you have not participated in a QFD activity yourself, identify someone who has, and create with that person a checklist of QFD planning categories. Compare it to the categories provided in this chapter. Reconcile the differences.

Using the QFD Estimator Chart, compare the actual time spent in a previous QFD project with the Chart's projections, and reconcile the differences.

[1] Quality Function Deployment: A Study of Implementation and Enhancements," Amitabh Pandey, MIT Master's Thesis, 1992.

[2] Thomas L. Saaty, "Decision Making for Leaders" and "The Analytical Hierarchy Process," both published by RWS Publications, 4922 Ellsworth Avenue, Pittsburgh, Pennsylvania 15213.

[3] Where information or estimates are available, quantify "high," "medium," and "low" with actual amounts. For example, "high" = $50M or higher, "moderate" = $5M to $49M, "low" = less than $5M.

[4] Thomas L. Saaty, "Decision Making for Leaders" and "The Analytical Hierarchy Process," both published by RWS Publications, 4922 Ellsworth Avenue, Pittsburgh, Pennsylvania 15213.

[5] "New Techniques for Listening to the Voice of the Customer," Robert Klein, Applied Marketing Sciences, Inc., in Transactions from the Second Symposium on Quality Function Deployment, June 18–19, 1990 (jointly sponsored by the American Society For Quality Control, American Supplier Institute, Inc., and GOAL/QPC).

[6] Shoji Shiba, Alan Graham, and David Walden, "A New American TQM: Four Practical Revolutions in Management," Productivity Press/Center For Quality Management, August 1993.

CHAPTER 16

Phase 1: Gathering the Voice of the Customer

Introduction

This chapter is a survey of methods for acquiring the Voice of the Customer. As we saw in Chapter 5, this is a complex subject. Listening to the customer requires time and good listening skills.

In this chapter, we'll first review the distinction between qualitative and quantitative customer needs data. Next we'll describe a method of classifying the customer needs into categories that are very similar to the way the Kano model classifies technical characteristics.

We'll then move on to a survey of some of the main methods for capturing the customer words. These methods include focus groups and one-on-one interviews. The one-on-one interviews will be further classified according to whether the interview is conducted at the customer's location (in context), or away from the customer's location. We'll draw enough distinctions between these types of interviews so that a development team could decide which methods to use.

We'll then discuss a completely different, but powerful, method for capturing the Voice of the Customer: analyzing customer complaints.

At the end of the chapter we'll present an overview of how all these methods fit together, along with a summary of a proven variant, the so-called "CQM" method of listening to the customer.

16.1   Qualitative Data and Quantitative Data

The QFD process requires that customer data be represented as a list of product or service attributes that are important to the customer. The attributes, or needs, are potential benefits that the customer could receive from the product or service. Each attribute in the list is to have some numerical data associated with it: relative importance of the attribute to the customer, and the customer's satisfaction performance level of similar products with respect to that attribute.

We call the attributes "qualitative" customer data, and we call the numerical information about each attribute "quantitative" data. Diagram 16-1 indicates where in the House of Quality these types of data go. The methods for arriving at the qualitative and quantitative data are different. In addition, there are many alternative methods for arriving at each kind of information.

A full discussion of the various methods for discovering the qualitative and quantitative customer data would be the topic for one or more books on market research. In fact, many QFD projects begin with complex and expensive market research projects. However, many product and process development teams cannot afford the luxury of hiring market research companies, and must therefore gather their customer data for QFD themselves. Whether they use a market research department or firm, or whether they do the work themselves, it is important for the development team to understand how the data acquisition is being done, and to understand what it means and what it *doesn't* mean.

If you think of the QFD process as a kind of machine, which accepts customer data as input and which produces a product or service as output, then it's clear that with faulty input, there's a substantially reduced chance for worthwhile output. Thus, the Voice of the Customer acquisition stage is crucial for the success of the entire endeavor.

In this section, we'll describe some important techniques that the development team must be aware of (whether they use the techniques themselves or others use them) in order to make informed decisions about their Voice of the Customer acquisition activity.

The general procedure for Voice of the Customer acquisition is:

> *first* determine the customer attributes (qualitative data);
> *then* measure the attributes (quantitative data).

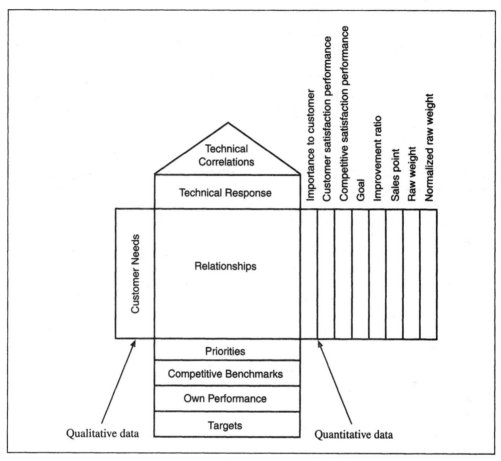

Diagram 16-1. Qualitative and Quantitative Data in the HOQ

Qualitative data is generally acquired by talking to and observing customers, while quantitative data is generally acquired by surveys or polls.

16.2   Classifying Customer Needs

It is possible to classify customer needs into categories that help development teams make decisions. While the Kano model is useful conceptually for understanding that customers have a range of responses to product *character-istics* based on their needs, we still need a model for classifying customer *needs* themselves. A model for this classification was developed by Robert Klein of Applied Marketing Science, Inc., to whom we are deeply indebted for assistance in explaining his model.[1]

Klein points out that there are two different ways to measure importance of customer wants and needs (attributes): by *directly asking* customers, or by *inferring* importance based on other data.

Directly asking involves one or another of the techniques to be described later in this chapter. All these techniques have one thing in common: the customer is asked how important a specific attribute is, either with or without reference to other attributes. Attribute importance measured by such direct methods is called *stated importance*.

A method for *inferring* importance is to measure how strongly satisfaction performance with an attribute is linked to overall product satisfaction. We might observe statistically (as in Diagram 16-3) that high levels of attribute satisfaction performance correlate to high levels of overall product satisfaction, while low levels of attribute satisfaction performance correlate with low levels of overall product satisfaction. We can then infer that the attribute is important, regardless of its stated importance. Using this same reasoning, an attribute whose satisfaction performance levels are not statistically linked to overall satisfaction (Diagram 16-2) can be inferred to be less important, unimportant, or possibly linked to importance indirectly. Attribute importance measured by this indirect method is called *revealed importance*.

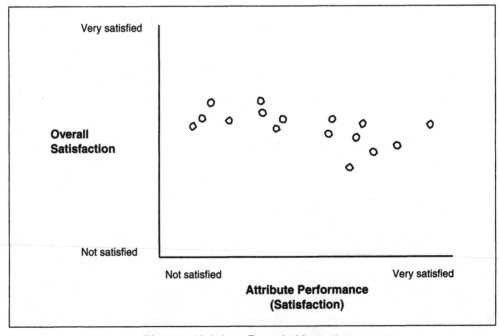

Diagram 16-2. Low Revealed Importance

Diagram 16-3. High Revealed Importance

In Diagram 16-2, we see considerable variation in customer satisfaction performance on a particular attribute, but very little variation with overall satisfaction. We can conclude that the satisfaction performance on this attribute is not a predictor or a determiner of overall satisfaction (or at least, is not *directly* a predictor). In other words, the revealed importance of this attribute is low.

In Diagram 16-3, the linkage between individual attribute satisfaction performance and overall satisfaction is quite clear. Customers for whom this need was well met were also satisfied with the overall product; customers for whom this need was not well met were also dissatisfied with the overall product. The conclusion is that satisfaction performance with this need is a strong predictor or determiner of overall satisfaction.

Depending on how attribute satisfaction performance is collected, the data can show a variety of possible relationship patterns. Interpretation of this data must take into account whether the customers surveyed were using one product or several, and whether the customer sample consisted of customers all of the same general type or of several market segments.

The main idea is to distinguish between customer attributes depending on how strongly they are linked to overall satisfaction. To do this we must measure both satisfaction performance on each attribute, and overall satisfaction with the product. We must then test for statistical correlation between these variables.

The Klein model now uses both the revealed importance and the stated importance of each attribute to classify customer needs into four useful categories:

Expected needs: High stated importance, low revealed importance
Low-Impact needs: Low stated importance, low revealed importance
High-Impact needs: High stated importance, high revealed importance
Hidden needs: Low stated importance, high revealed importance

16.2.1 Expected Needs

Expected needs are those basic needs that customers insist must be met. If the needs are not met, customers will be strongly dissatisfied. If the needs are met, customers will be only moderately satisfied with the product. Expected needs are very similar to "Expected Quality" in Kano's model.

An example of an Expected customer need for breakfast cereals might be

Comes in the size package I need

For this need, customers may be very dissatisfied with the product if they cannot get it in the package size they need, but will probably not be highly satisfied with the breakfast cereal just because they are able to buy it in the package size they need.

16.2.2 Low-Impact Needs

Low-impact needs are needs that customers may articulate, but which in practice have no relationship, or at least no direct relationship, to the customer's overall satisfaction with the product or service.

An example of a low-impact need would be for breakfast cereals with

Recipes for innovative new uses of this product

People might buy the product regardless of whether recipes are made available. (In the case of breakfast cereals, buying patterns are almost synonymous with satisfaction.)

There is no construct in the Kano model having a counterpart with low-impact needs in Klein's model.

The reader may ask: how could a low-impact need ever emerge as a need at all? This could happen if interviewed customers are aware of the need being addressed by some product or service providers, especially if the way the need is addressed is different with different providers.

16.2.3 High-Impact Needs

High-impact needs are those that cause high customer satisfaction when they are met, and low satisfaction when they are not met. These are similar to the customer needs for which Kano's "Satisfiers" or "straight-line quality" are the solutions.

A high-impact customer need for breakfast cereal might be

Cereal is nutritious

People may prefer nutritious cereals, and may exhibit that preference by buying such cereals. If so, the need is high-impact, because cereals with lower nutritional value would not be purchased as often—that is, customers would be less satisfied with such cereals.

16.2.4 Hidden Needs

Hidden needs are those that customers say are not important to them, or that customers do not mention, but that, if met, strongly affect customer satisfaction. To identify these would obviously provide a product or service developer with a competitive advantage. These are the needs for which Kano's "Delighters" are the solutions. In general, examples of "Delighters" given in explanations of the Kano model are technical characteristics that meet unspecified needs, rather than needs themselves.

A well-known case from market research relates to breakfast cereals containing sweeteners. Most people *say* that sweeteners are not important to them. This would be represented by the customer need

No sweeteners in my cereal

Yet, their buying habits, and therefore their satisfaction with such breakfast cereals, contradicts that statement—sweetened breakfast cereals generally outsell unsweetened ones. Thus, we have hidden customer need: stated importance is low, yet revealed importance is high.

16.2.5 Klein Grid of Customer Needs

A simple grid shows the relationships between stated importance and revealed importance, and how they relate to the four needs categories (Diagram 16-4).

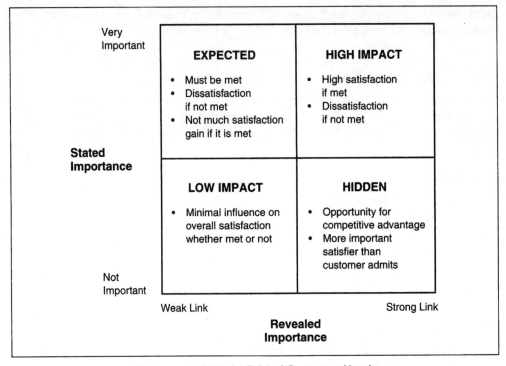

Diagram 16-4. Klein Grid of Customer Needs

Knowing whether customer needs fall into one or another of the four Klein categories can help development teams to set customer satisfaction performance goals and determine sales points. Generally, goals would be set high for attributes classified as High Impact and Expected. Attributes classified as Low Impact generally don't warrant high satisfaction performance goals (unless there is some indirect linkage to attributes in other categories.) Attributes classified as Hidden must be looked at carefully to determine the strategic impact of goal setting.

16.3   Gathering Qualitative Data

16.3.1   Customer Needs and Technical Solutions

One of the most important, yet difficult, distinctions that product developers must make is between customers' needs, on the one hand, and technical solutions to meet those needs, on the other. Let's take a few examples to illustrate this.

A person shopping for an automobile may ask for a car with tinted glass. Is this a customer need? In all likelihood, no one actually *needs* tinted glass. Instead, people need some of the things that tinted glass may provide. Tinted glass can reduce glare from bright light, such as direct sunlight, or headlights of other cars. Tinted glass can also provide privacy to passengers, by making it difficult or impossible for people outside the car to peer in. Tinted glass can also help to keep the auto interior cool, by reducing infrared radiation from the sun and from warm objects near the car.

For each of these possible needs—reduced glare, privacy, and cooler interior—a variety of technical solutions are available, some of which may be more effective or less expensive than tinted glass. Glare can be reduced by altering the shapes and finishes of surfaces that reflect glare, such as the dashboard or the hood. Privacy can be achieved by custom-fitted venetian blinds or curtains, or by no windows at all. A cool auto interior can be achieved by ventilation and air-conditioning methods.

In an example from the world of computers and software, central processor speed is often expressed in "MIPS," millions of instructions executed per second. Another measure of processor speed is "clock speed," usually expressed in megahertz, or millions of clock ticks per second. Some technically oriented computer buyers will specify processors according to their speed rating in MIPS or megahertz.

But it's the rare computer user who actually needs "20 MIPS" or "25 megahertz clock speed." Instead, they need computers that perform certain tasks within certain time frames. The tasks could be to rapidly recompute the contents of a spreadsheet of a certain size, or to print or display complex graphics rapidly. The speed at which these tasks are performed is to some extent related to the processor speed, but most computer tasks also depend on other technical factors, such as main memory capacity (usually referred to as RAM), input–output capacity, floating-point speed, and other factors. Depending on the real need of the computer user, as well as on the design of the software that performs these tasks on the computer, these other factors may actually have a greater influence on performing the user's tasks than does MIPS or clock speed.

In order to make design decisions that meet customers' needs, product developers must understand the customers' real needs, often buried beneath their surface-level requests. Product developers must be able to distinguish between customers' real needs and the technical solutions to those needs. Just as customers don't make these distinctions, many product developers have not seen the necessity to make such distinctions, either.

The ability to make clear distinctions between needs and solutions is a prerequisite to generating breakthrough concepts. The original Sony Walkman met a customer need that had hitherto been poorly met at best: the need for portable, high-quality playback of stereo music. This is not to say that by simply distinguishing between customer needs and technical solutions, a product developer is assured of a breakthrough concept. However, the habit of not accepting technical solutions as customer needs will certainly help product developers to find creative solutions to delighting customers.

The QFD process puts customer needs on the left side of the House of Quality matrix, and technical solutions on the top of the matrix (Substitute Quality Characteristics). If these get mixed up, filling in the House of Quality will be very confusing. In fact, when people get confused during QFD, I often suggest we check whether needs and solutions may be mixed up. It's essential for a smooth QFD process, as well as for a successful product or service.

The question that product developers must answer before they can design their products is: *Why does the customer want the technical solution being asked for* (tinted glass, or 25 megahertz central processor speed)?

Very often when customers tell us what they want in a product, they will ask for what they believe is the best technical solution to their unstated need. However, customers are not necessarily the most technically knowledgeable people with regard to the design of your product or service. Thus, there is no a priori reason why they should possess the best solution. The product developer is usually better qualified to identify effective technical solutions to meet customer needs, but he or she must have a clear idea of those needs in order to do so.

Many product development engineers will say that customers don't really know what they want. What probably supports that assumption is that many customers tend to ask for inappropriate technical solutions, rather than state what they want.

In fact, customers are the experts in what they need, but they are rarely given the opportunity to discuss their needs with product designers. One reason for this is that product and process designers often don't have the time to discuss customer needs with customers. When they do, they don't necessarily ask the right questions.

For example, asking the direct question "What do you want?" is an invitation to a customer to provide a technical solution. And yet this is a very natural question for a product designer to ask a customer. The important follow-up question, "Why do you want that?" is not one that product developers have been trained to ask.

A skilled interviewer uses "probing" questions as a tool to uncover the unstated needs of the customer. Useful variants to "Why do you want that?" are

- If you had that, what would it do for you?
- Have you ever had that? How did it work out?
- What's good about that?

Another key is to listen to the words and phrases that should invite a probing follow-up question. Customer terms such as "good," "bad," "easier," "don't like/do like" are invitations to the alert interviewer to probe more deeply to find the real customer wants and needs.

The most important thing the interviewer must do during customer interviews is to persist in pursuing the customer needs by probing beyond the surface responses of the customer. Then, when analyzing the customer's responses, make clear judgments as to which customer responses represent needs and which represent solutions.

16.3.2 Focus Group Interviews

The focus group process involves assembling a group of customers (respondents) together in a room, and facilitating a discussion in which each respondent states his or her views to the group and can hear and respond to the other group members' comments. Focus group interviews are so named because the group is directed in its discussion to "focus" on certain topics determined by the group facilitator.

The number of respondents in a focus group is generally between five and fifteen. The larger the group, the more skillful must be the facilitator in order to keep the discussion on the desired topics.

The most commonly stated advantages of focus groups are

Synergy—one group member's comments may prompt another group member to discuss a topic no one would otherwise have thought of.

Cost and efficiency—several customers are interviewed in the time it would otherwise have taken to interview only one customer.

Some generally acknowledged drawbacks to focus groups are

Tendency toward intersubjectivity—that is, an emphasis on those beliefs that the group members share—at the expense of intrasubjectivity, an emphasis on those beliefs specific to a few but not all of the group members.

Lack of adequate "airtime" for any one group member, since the available time must be shared by all group members.

The goal of a focus group interview is the same as for most other forms of interview: probe beneath the interviewee's surface-level comments to arrive at the true needs. Many of the same considerations mentioned elsewhere about interviewing style apply to focus groups.

Additional interviewer's techniques are also important. The interviewer must act as a facilitator; that is, he or she must control the interactions among the group members, as well as the interactions between each group member and the interviewer. One group member may contradict another, and the one being contradicted may take offense and seek to defend his or her comment. Such a discussion is not likely to generate additional customer wants and needs, and must therefore be controlled by the facilitator.

Some group members may be voluble, others may be reserved, and yet the objective is to capture all of their needs. The interviewer must find ways to encourage the talkative ones to quiet down, and the reticent ones to speak up.

The possibility for two or more people to speak at the same time creates difficulties for the notetaker or transcriber, so the facilitator must exercise a kind of "traffic control" function, aimed at allowing only one person to speak at a time.

There are special circumstances where the focus group approach may be the only possible way to collect customer wants and needs. For example, the peer relationships between group members may dictate that the focus group style should be used. The group members may be more comfortable speaking to each other than speaking to an interviewer in a one-on-one setting. Consider a group of physicians, who may share a common vocabulary and attitude toward their profession that makes them unwilling to discuss their feelings and beliefs with an interviewer whom they may regard as a layman.

Generally, focus group interviews are conducted in specially constructed rooms. These rooms contain a table, around which the interviewees are seated. The facilitator may sit at the head of the table or in some other position that indicates he or she has a special role to play during the interview. At one end of most focus group rooms, the wall has been fitted with a large window of coated glass, the so-called "one-way" mirror. Behind the mirror, and therefore out of sight of the interviewees, there may be a video camera and a tape recorder, and also several people who wish to observe the interview but who want to otherwise influence it as little as possible. Most interviewees, or respondents, as marketing specialists like to call them, become accustomed to the fact that they are being watched and recorded, and seem not to be intimidated by that fact.

16.3.3 One-on-One Interviews and Contextual Inquiry

Conference Room Interviews

The most common form of one-on-one interview takes place in a conference room or office, sometimes in a neutral area (neither the interviewer's nor the interviewee's normal location). Depending on the availability and ease of scheduling interviewees, either the interviewer travels to the interviewee's site, or the interviewees travel to the interviewer, who stays fixed in one location. The latter is cheaper and more efficient for the interviewer, but is not always possible.

The interviewer will have prepared some sort of "interview guide" that acts as a reminder about what topics to discuss during the interview. The typical interview guide indicates the following:

- Important points for the interviewer to make when beginning the interview, such as explaining the reason for the interview, explaining the style of interview to be conducted, and asking permission to audiotape or videotape the interview
- Suggested interview opening question (usually a general open-ended question)
- List of special topics that should be addressed during the interview
- Closing questions and comments

The interviewee will have been selected according to screening criteria that ensure the type of experience, knowledge, attitude, or other characteristics that are needed to discuss the topics that the interviewer wants to cover. Many market research groups have developed interviewee recruiting and screening techniques to ensure that the interviewees can provide the information that the development team needs.

Generally the interviews are recorded, either on audiotape or videotape (with the interviewee's permission). However, because of the cost of transcribing or even of reviewing the recording, some interviewers prefer to skip the recording process and simply take notes. In my experience, it's well worth the money and time to record, transcribe, and analyze the transcript. There are two reasons for this:

1. If you know you will be analyzing a transcript, you don't need to take notes. In that case, you can concentrate fully on what the interviewee is saying, and you will be more able to detect important points that the interviewee is making as they are being made, so you can follow them up instantaneously.

2. Very often, the interviewee will tell the interviewer something very important at the beginning of an interview, but will use vocabulary or

jargon that the interviewer will not yet have learned. In this case, the interviewer may not understand the importance of what was just said. An interviewer taking notes will probably miss this important point completely, because the jargon will not make any sense. On the other hand, the point will appear in the transcript, because the audio or videotape records everything that occurs without regard to whether it makes sense.

If the interview is recorded, the transcript can be analyzed after the entire interview has been conducted, at which time the interviewer has had a chance to learn the vocabulary or jargon of the interviewee. In this case there is a much better chance that the interviewer will understand the significance of the comment.

Open-Ended Questions

Open-ended questioning is a critical technique during the interview. The term open-ended questioning refers to questions for which there is no simple *yes* or *no* answer, and for which there is no right answer, nor could the interviewee guess at a right answer. Such questions encourage the interviewee to provide much more information than a simple "yes" or "no" could provide.

In contrast, so-called "leading questions" are the exact opposite of open-ended questions. They invite the interviewee to provide a short answer that adds little or no information to the discussion. Leading questions often suggest an answer, or give the interviewee very little choice in answering. Leading questions are often designed to help the interviewer prove a point, rather than gain new information. The interviewer is typically not interested in gaining new information, but in confirming an already-held belief. The following are examples of leading questions:

"Are you satisfied with this product?" (The interviewee may answer "Yes" out of politeness or because he or she is satisfied. Even if the answer is "No," very little information has been provided. In any case the interviewee has very little room to provide a more helpful answer.)

"Do you use a seatbelt when you drive?" (Most people are aware that seatbelts *should* be worn. Hence, they are likely to respond with the "correct" answer that they use seatbelts even if they don't.)

Let's turn the same questions into open-ended questions.

"Are you satisfied with this product?" becomes "What have your experiences been with this product?"

"Do you use a seatbelt when you drive?" becomes "What are the steps you take in starting your car?"

In both these open-ended questions, you have encouraged the interviewee to provide much more information than with the leading questions. Also, you have avoided suggesting answers. Diagram 16-5 summarizes some differences between open-ended questions and leading questions.

The best interviews are those in which the interviewer strikes a balance between restricting the interview to just those topics covered by the interview guide, and letting the interviewee discuss any topic under the sun. In the first case, the interviewer is sure to get responses to the topics identified in advance as important, but quite possibly no others. In the second case, the interviewer will probably hear about many topics that could not have been anticipated, but may not hear about some topics identified in advance as important.

It's usually a good idea to give the interviewee as much freedom as possible during the early part of the interview. During this period, the interviewee will have an opportunity to touch on unanticipated topics, and also—without explicit prompting—on some of the anticipated topics. In the latter part of the interview, the interviewer can exercise more control. He or she can lead the interview to those topics from the interview guide that have not already been covered, or choose to focus on the unanticipated topics that came up in the first part of the interview.

An important rule to bear in mind in interviews: if you've learned about a customer want or need just once, from a single customer, you don't need to hear about it again. The purpose of interviews is to uncover all possible wants and needs, without regard to their relative importance to the market as a whole. The quantitative part of the Voice of the Customer acquisition process will distinguish between the more important and less important wants and needs.

Focus

What we look for determines what we see. The term for this phenomenon is *focus*. Interviewing experts understand that we always have a focus, whether we are conscious of it or not. In order to learn what we need to learn from any interview, we must notice what our focus is, and if necessary, redirect it to what best serves our interest.

What focus does. Try this experiment with two colleagues in your office. Ask one colleague to wander around the office and make a note of all office

Leading Questions	Open-Ended Questions
Are you satisfied with this product?	What have your experiences been with this product?
Do you use a seatbelt when you drive?	What are the steps you take in starting your car?

Diagram 16-5. Leading Questions and Open-Ended Questions

furnishings he sees. Ask another to wander around the same area, and write down all papers she sees.

Now compare the two lists. Not surprisingly, they will be quite different, even though each person was looking at the same area at the same time. The difference is due mainly to what they were focusing on. Other sources of difference will be due to their different personalities and to how each person understood the task. But the prime difference is because of focus.

We have choices about setting focus. We can choose to look at a wide range of things, or to concentrate on just one or two aspects. We can choose to focus on people to the exclusion of machines, or on processes to the exclusion of results. Whatever we choose, and whatever we focus on, we will see what we focus on, and we will not see what's outside of our focus (Diagram 16-6). Thus, since our choice of focus both reveals and conceals, we must manage it.

Focus helps us decide what questions to ask, and what information to seek. It determines what we will understand, out of all the data we are bombarded with.

What makes focus vary. While it's important for us to be aware of and to control our focus, unexpected events often take over. New data received during an interview may cause us to reevaluate some basic assumption.

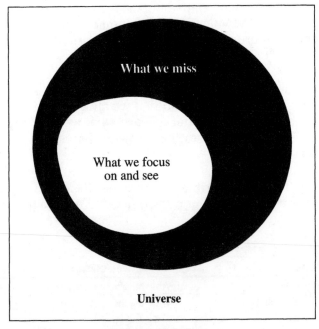

Diagram 16-6. Focus

For example, say we want to understand our customer's perceptions about the reliability of our product. We may have chosen to focus on the implications of failure rate on our customer's daily routine. After discussion with our customer, we may discover that the failure *rate* is not the customer's main concern. Instead, it's our service technician's response when a failure occurs that is much more important to the customer. Obviously, our original focus would not have been on the mark.

This type of surprise occurs frequently during interviews. The fact that open-ended interviewing uncovers unexpected areas of focus is one of its strengths. In order to take advantage of such surprises, we must be prepared to alter our focus.

Sometimes it will not be possible to alter the focus on the spot—we may suddenly find ourselves interviewing the wrong person, or conducting the interview in the wrong location. In such a case, our best approach is to replan our objectives for later interviews.

Another way to create the possibility for focus to vary is to share your interpretations of what you are observing with the customer. Your understanding may be just slightly different than what the customer was intending to explain. When you offer your interpretation, the customer has the chance to correct any misunderstandings. Should your misunderstanding be significant, you will of course be able to adjust your view, and you may have to change your focus.

As you continue your inquiry with the customer, you may develop some design ideas. It's a good idea to share these with the customer. The customer may be able to confirm the validity of your ideas, or show you in what way you're off the mark. Again, this can lead to change of focus.

In summary, recognize that your focus both reveals and conceals what you will understand during an interview. Be prepared to interrupt the interview to ask for concrete evidence of what is not visible, to share your interpretations with the customer, and to share your design ideas with the customer. Don't wait too long to create these opportunities for refocusing. The longer you wait, the more danger there is that you will have focused on the wrong things and missed important data.

Planning the Interviews
Interviews, and especially their analysis, take time. It's a good idea to plan your work so the time is well spent. Here are some recommended steps.

1. Identify the questions you want answered. A useful method for identifying these questions is to have the development team brainstorm all questions and then affinitize them. If the number of questions is very large, you may wish to prioritize them, using either the Prioritization Matrix method or the Analytical Hierarchy process. These questions won't be asked outright of the customers, unless by the end of the normal interview some of them have not been addressed.

2. Identify your target customers. Once the questions are known, you'll have a better idea of how you want to segment your market, and what types of companies or individual users you wish to interview.

3. Select participants. In many cases, as we indicated in our discussion of key customer groups earlier, the end user of your product may not be the only person who must be satisfied. For example, a parent makes buying decisions for children, and an office manager makes buying decisions for secretaries. In this case, it's a good idea to conduct interviews with people representing all the important roles in the selection, installation, use, and servicing of your product.

Sometimes a matrix[2] may be useful for determining how to get the coverage you need in the interviews. Try putting the market segments along the left and the roles along the top. In Diagram 16-7 we have five market segments and three roles identified. If we don't want to spend the resources for all fifteen interviews, we can distribute our interviews across the columns and rows judiciously, and design a type of coverage that will be the best compromise. In this example, we've minimized the number of interviews in Europe, while also ensuring pretty good coverage of "Users" over three of our five market segments. Also, at least one of each type of "User" and at least one of each market will receive an interview. The total number of interviews is six, versus the potential of fifteen.

4. Establish the right environment for the interview. Make sure that whoever recruits and screens the interviewees explains to them what the general topic is, how long the interview will take, and when and where it will occur.

Market Segments	Purchaser	User	Maintainer
Northeast U.S.	Interview		
South U.S.		Interview	
West U.S.		Interview	Interview
Latin America		Interview	
Europe	Interview		

Diagram 16-7. Covering the Segments/Users Grid

Get permission in advance to record the interview, to avoid having the interviewee balk at the last moment.

5. Set the focus. A final step in planning the interviews is to explicitly decide on the focus. The developers will normally plan several interviews at once. The planned focus may differ for each interview, and that in turn may help the developers decide who should conduct each interview—the reliability expert might wish to speak to the maintainer, for example.

6. Replan after every interview. Regardless of all previous planning, experience has shown that the interviewers will want to replan their interviews after each interview has been completed. This is because each interview will bring a number of discoveries, and these in turn will modify the goals of subsequent interviews. Be sure to review the results of each interview as soon as it is over, and be ready to modify your plans based on the results.

Telephone Interviews

Telephone interviews are inexpensive compared to other styles of interviews. The major cost-saving element is the absence of travel expenses. The interview itself is also often shorter than face-to-face interviews, although there is no physical or technical reason, only a psychological reason, why this should be so. The recruiting and screening process for telephone interviewees is the same as recruiting for face-to-face interviews.

Other important similarities with face-to-face interviews:

- The interviewer can (and should) prepare and work from an interview guide.
- The interviewer can, after receiving the interviewee's permission, record the interview.
- The general strategy of asking open-ended questions, and of providing the interviewee freedom to discuss any topics, is still the best strategy.

The main differences between telephone interviews and face-to-face interviews are these:

- The technical quality of a telephone interview recording is rarely as clear as that of a face-to-face interview, because the dynamic range and frequency range of voice of a telephone is limited. Nevertheless, inexpensive equipment is readily available for recording voice on tape.
- The visual clues of facial expressions and body language are not available to clarify to the interviewer what the interviewee is attempting to express.
- The same visual clues are not available to guide the interviewer as to when to switch to a new topic, when to probe, or when to terminate the interview.

- It is not possible over the phone to show the interviewee product samples or simulations, nor to observe the interviewee using or reacting to such samples.

Contextual Inquiry

Contextual Inquiry [3,4] is the name of style of interview that emerged in the early nineties as a technique aimed at helping developers to uncover hard-to-identify customer needs, and also to generate innovative solutions to meet those needs.

In one sense, Contextual Inquiry is a variation on the one-on-one interview method described above. Almost all of the principles that apply to one-on-one interviews apply to Contextual Inquiry. The difference lies in the additional stress on observing the customer *in the context* of the customer's activities, and on the interactive nature of the relationship between interviewer and interviewee—so interactive that it is called *inquiry* as opposed to *interviewing*.

Summary Data versus Concrete Data

Suppose you go to a party. The room is crowded, everyone is talking. Music is playing in the background. In fact, you might wish it to be further in the background than it is, because you can barely hear yourself speak. A stranger strikes up a conversation with you.

"What do you do?" the stranger asks.

Each of us no doubt has a different reaction to this common question. Some people don't like to talk about their careers at parties. Others are happy to discuss their work without limit. You have decided to provide a polite answer. You want to be friendly and informative, but because you are at a party, you don't want to limit your participation in the party to a discussion of what you do.

You could answer the question in many ways. After a moment's thought, you say:

"I'm an engineer."

You have just provided *summary data* to your new acquaintance. We call it summary data because it summarizes a vast amount of information (in this case, about what you do) into a very few words that you have tailored for this situation.

Summary data. What does your acquaintance now know, based on your answer?

First, what you don't do. For example, you don't remove appendixes in a hospital operating room. You don't drive a truck.

Second, that you do some sort of work—you're probably not unemployed or independently wealthy.

Next, that your work is somehow technical in nature.

There are obviously many ambiguities in your answer. The first relates to which engineering discipline you are involved with. Perhaps your work involves machines, electronics, roads, or aerospace, since there are many types of engineers. Even within these disciplines, the possibilities for erroneous assumptions is enormous. Some mechanical engineers are involved with auto engines, others with hand staplers. Some electronics engineers work with microchips, others with telephone receivers.

The second ambiguity relates to the kind of work you do within your engineering discipline. Some engineers design products. Others develop manufacturing processes. Others manage engineers and engineering operations, and don't actually do technical work themselves.

There are endless possible interpretations your acquaintance may apply to your answer. To what extent your acquaintance will truly understand "what you do" will depend on how much additional communication occurs between you.

Concrete data. Given the context of the party, your simple summary answer may have been appropriate. But suppose you had wanted your acquaintance to know as much as possible about what you do. How else might you have responded?

A completely different, albeit extreme, approach would have been to invite your acquaintance to gather *concrete* data. You might ask him to meet you at your home the following Monday morning and "shadow" you at your work every day thereafter for the next two months. He would observe your commuting experience, your office; he would learn your secretary's name, what type of coffee you drink, and who your colleagues are. He would see how meetings are conducted, what type of phone calls you receive, what mail you read, what mail you throw out, how you communicate to your customers, how much time you spend at your desk, and what you do while you are there. He would learn whether you conduct work while commuting—for example, by cellular phone—whether you bring work home, how much time you spend away from the office on your work.

Because your acquaintance would have followed you for two months, he would understand some of the cycles in your work. Your acquaintance would have observed periodic staff meetings and their occasional cancellations, as well as follow-up and lack of follow-up from these meetings. He would have seen priorities shifted, emergencies dealt with, deadlines met and missed. He would have observed in passing the effects of people missing from the office, on vacation, on business trips, or ill. He might have seen you absent yourself from the office, to attend a meeting with customers, to attend a training class, to take a vacation.

The ambiguities associated with the summary data received at the party would now be much reduced. By observing the building you work in, the conversations you have, and the mail you receive and send, your acquaintance would know precisely which engineering discipline you are involved with. This knowledge would be bolstered by pictures, models, or actual machines, so that many specifics, such as your current project and your current schedule, would be revealed. It would include information on whether you manage others, and if you do, what your management style is. If you are an individual contributor doing technical work, he would see by example how you do your work: at a computer, at a drawing table, or in a laboratory. He would know what part of your work you do alone, what part in communication with others, and by what types of communication (by phone, in person, by electronic mail, by fax, or at meetings large or small, short or long).

Not only would this in-depth observation avoid ambiguities, but it would reveal the *affect* of your work as well. In other words, your mood and the moods of the people with whom you work would become evident. Your acquaintance would see moods shift, by the hour and by the week. He would see arguments, celebrations, smiles, and frowns. The *pace* of the work, its rhythm and its urgency would become visible. He would understand what parts of the work are repetitive and what are the lengths of the cycles.

All these observations would add to a far richer understanding of "what you do." Your acquaintance would experience your work as deeply as possible, short of actually taking over your job and doing it for you.

This very long description of the observations your acquaintance would make is an illustration of the rich knowledge we can get from direct observation. This type of detailed, fully immersed observation characterizes one of the differences between conference room interviews and market research in its most common form, and Contextual Inquiry.

One fundamental difference between data gathered at a party or in a conference room, and data gathered by direct observation, is the *concreteness* of the directly observed data. By experiencing things concretely, we learn *more* than can possibly be written down. By learning of things in summary form, we learn *only* what can be written down.

Shared Experience: Another Kind of Reality

One of the main principles behind Contextual Inquiry is that the fullest understanding we can possibly get of our customer's needs is obtained by not just observing, but actually *sharing* in the customer's experience as fully as possible. This shared experience helps us to appreciate the customer's needs almost as deeply as if the needs were our own.

With Contextual Inquiry, this means observing the customer living his life as it impinges on our product or service. If our customer uses our product at work, then we must share that work experience with the customer. If our customer uses our product at home, then we must share the home experience with the customer.

What the Engineer Can Share with the Customer. The notion of "sharing" experiences with strangers might seem to imply more intimacy than many of us are willing to contemplate.

In practical terms, the sharing is no more than a professional relationship aimed at providing the product or service developer with design data. It includes observation of the customer's activity, at the site of that activity. It also includes discussion with the customer to fully appreciate how the customer experiences that activity. If the product is a leisure product, it involves observation of the customer's leisure as it applies to that product, at the site of that leisure.

What Is Design Data?

By design data, we mean information that provides the developer with information that can be used to make the product or service clearly support the customer's goals. Thus, the requirement "easy to use" would not be design data, since we have no information on what aspects of use are important to the customer.

What might be closer to design data would be the requirement "easy to correct my mistakes." If we had seen our customer make some mistakes, and then attempt to correct them, we would understand what the obstacles are, and what about our product might remove those obstacles.

When we seek design data, we have three objectives in mind for our product or service. For whatever are the customer's applicable activities, we want to **support** the customer's needs in pursuing those activities, we want to **extend** his or her pursuit of those activities, and we want to **transform** the customer's pursuit to some level or domain that not even the customer thought was possible.

An example of these three objectives will help clarify the distinctions. Before 1975 or so, the primary device for creating legible documents in the office was the typewriter. The typist's activity was aimed at creating a neat, legible document quickly. The most frequent disruption to this activity was probably the typing error. When a typist made a typing mistake, he or she had to stop typing, roll the paper back so the typing error was within reach, erase the mistyped material with a special abrasive eraser, roll the paper back in place, and retype the material correctly.

With the advent of word processors, the typist's work has become **supported** in this area by the "delete key." Now when a typing error occurs, the process of correction is much simpler—just hit the delete key one or more times until all the incorrect typing is removed, and then continue. The issues of overzealous erasure (rubbing a hole in the paper!), or of misaligning the paper when rolling it back, have been made obsolete with the word processor. Thus, correcting a typing mistake with a word processor is far less disruptive than with a typewriter.

Word processors have done more than support error correction. They provide pagination capabilities, so that page numbering, top and bottom margins, and headers and footers are automatic. They provide easily adjustable tabulation. Some word processors have made it easy to type special characters, such as mathematical symbols or foreign language characters. All of these capabilities have **extended** the typist's work far beyond what a typewriter could readily support.

In both these examples, *support* and *extend*, the underlying paradigm is still the same. First, we compose the document—we generally write it in longhand, for instance. Then, we take our material to the typewriter or word processor and transcribe it.

Word processors since the late eighties have gone much further and opened the possibility for a completely new paradigm. By providing rapid "cut and paste," integrated outlining capability, spell checking and grammar checking, they have **transformed** the process of creating a legible document. Many people now *simultaneously* compose and transcribe their documents. This is a process that was all but impossible in the days of the typewriter. Granted, many people still prefer to write documents in longhand, and then type them. But many others, and a growing number of them, now prefer the transformed process of simultaneous composition and transcription.

To summarize, in Contextual Inquiry, we attempt to get design data that shows us how to support the customer's current activity, how to extend the activity, and finally how to transform the activity into something considerably more useful or attractive.

Accepting the Impossible

To be open to paradigm shifts that can lead to transforming the customer's activity, product and service developers must temporarily accept that anything is possible and nothing is impossible. In order to shift customer's paradigms, we have to shift our own paradigms. Paradigms are determined by our implicit beliefs about what is and what is not, and about what can be and what cannot be.

When listening to the Voice of the Customer, it is the customer's unreasonable, impossible demands that have the potential to help us break out of our current paradigms. Our job is to hear the demands as expressions of the customer's needs. Our job as product or service developers is to use our special technical knowledge to help meet the customer's needs. The more open we are to accepting all the customer's needs as possible needs we can meet, the more likely it is that we will conceive of a product or service capability that will meet a customer need that the competition hasn't understood or acknowledged.

Contextual Inquiry: Context

As indicated in the scenario of "shadowing" an engineer to understand his or her work, there are many aspects to the context of the customer's use of our product. Some of these aspects of context play a major role in how the customer will use our product. Our job as a developer is to identify these key contextual aspects.

No one can know what all of these contextual aspects will be, or which to pay attention to. We can only be aware of the many possible aspects and be ready and willing to observe them when we visit the customer at the site of product use. Probably the most important aspects of the user's context are the ones that differ greatly from the developer's. Those will be less familiar to the developer, and also most difficult to simulate in a laboratory.

Aspects of the user's context. Some of the more important aspects of the user's context are location, people, culture, and values.

Location can be divided into immediate vicinity and more remote areas. The immediate location could be indoors or outdoors.

If indoors, is it in a cramped or spacious enclosure, is it hot or cold, noisy or quiet, humid or dry? Is air quality important? Are safety, comfort, or privacy important issues? Do things happen quickly or slowly? When using the product or service, is the customer lying down, sitting, or standing up? Is the area brightly lit or very dark? Over what period of time does the use occur—for five seconds at a time, ten times per day, or twenty-four hours per day, every day?

If outdoors, then the immediate location includes whether the use is under water, at ground level, or above ground. What type of weather and climate prevail? Is the product used only during certain seasons, or year round? Is the product used at night, or during the day? Is the product in a fixed position, or is it expected to move? Is air quality important? What type of clothing is the customer wearing—gloves, sunglasses, safety goggles, sneakers, steel-toed shoes?

The number of possibilities is limitless. This is why it's so important for the product or service developer to actually experience the places the product is used, rather than make assumptions.

People include those who are in the immediate vicinity of the customer, and those who are not nearby, but who have important relationships to the customer. Either at a distance or nearby may be co-workers, colleagues, friends, students, teachers, patients, clients, teammates, or opponents. Other relationships of people to the customer that may also be important are boss, subordinate, parent, child, sibling, customer, supplier, coach, enemy, or lover.

Culture covers the issue of how people relate to each other and to the customer. It includes people's beliefs, customs, methods, and styles of communication, decision-making processes, and organizational structures. For each of these aspects of culture, we may need to know what happens between people, when things happen, how they happen, who is involved, and why? All these aspects of culture may have important implications for how our product or service will be used by the customer.

One of the most important aspects of culture that product and process developers overlook is the customer's language. All too often developers describe their creations in terms that have technical meanings that the customer does not understand. Likewise, customers will refer to their activities using words that have no meaning to the developer. Under these circumstances it's unlikely that the developer's product or service will be right on the mark. The developer has a responsibility to learn the customer's vocabulary as it pertains to the product or service under design.

Values relate to what is important to the customer. Aspects of values that may be important are what the customer respects, disrespects, enjoys, hates, tolerates, notices, and ignores. Most important to the developer is what represents *success* and *failure* to the customer. In a sense, the developer's job is to enhance the customer's chances for success in whatever the customer is doing that involves the developer's product or process.

Contextual Inquiry: Inquiry
When Contextual Inquiry is described to most people, they quickly grasp the significance of experiencing the customer's context. But they tend to ignore

the other key word—inquiry. Inquiry is very different from interview, and we'll try to draw the important distinctions here.

Key concepts of inquiry. The important image of inquiry in this discussion is one of partnership. As developers, we have special knowledge of our products and of the technology we use to develop our products. The customer has special knowledge of his or her job or activity and how our product fits in.

In a very real sense, we are both experts, but in very different subjects. Our job as partners is to create a large enough overlap in our areas of expertise so that our product or service will strongly support, extend, and enhance our customer's activity (Diagram 16-8).

Inquiry versus interview. When we approach a customer for an interview, both we and the customer conspire to take on roles that work against achieving the full level of understanding we seek. As interviewer, "I" ask the questions. As interviewee, "you" provide the answers. "I" am in charge of the direction of our discussion, and the range of topics we pursue. "You" are inclined to take a more passive role and simply respond to my questions.

In contrast, the inquiry can be seen as a joint search for information or for the truth.

In the inquiry model, there is little or no difference between our roles. We are inquiring together into the problem of how our product or service can help the customer achieve his goals. We both need to learn about the other's expertise, and we both must "try out" as many design possibilities as we can in our inquiry. Just as the classical market research interview intentionally separates

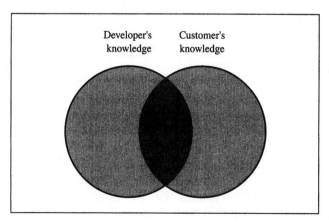

Diagram 16-8. Developer/Customer Partnership

customer needs from technical solutions, the Contextual Inquiry tries to create technical solutions from the customer needs.

In this relationship, we are striving for a partnership of developer and customer, in which the two together create the product or service concept.

We have a few simple tools that help us structure our discussions as inquiries rather than as interviews: stage setting, and open-ended questioning.

Stage setting is the process of enrolling the customer in the partnership. This involves explaining to the customer beforehand the main ideas of Contextual Inquiry, with special emphasis on the notion of creating together the ideas that will make the product serve the customer's needs. Most customers have been very willing to team up with developers in Contextual Inquiry. Some customers may hesitate because they feel they lack technical competence in product design. Here the developer should reassure the customer that the customer's expertise is an essential ingredient of the Inquiry; it is unique and irreplaceable.

As with all other forms of interviewing we have described in this book, open-ended questions create the best environment for Inquiry.

Planning the Interviews
Contextual Inquiry interviews, even more than other forms of interview, take time. The planning outlined for one-on-one interviews applies for Contextual Inquiry as well.

1. Identify the questions you want answered. Use the same techniques described for one-on-one interviews.

2. Identify your target customers. Use the same techniques described for one-on-one interviews.

3. Select participants. Use the same techniques described for one-on-one interviews.

4. Establish the right environment for Contextual Inquiry. Very often Contextual Interviews will be set up by account representatives, salespeople, or others who may be in direct contact with our interviewees, but who may not be familiar with the principles of Contextual Inquiry. These representatives may be very successful in getting our users to agree to be interviewed, but the users may still not know what to expect for the actual interview. They may not fully appreciate that the interviewer will want to conduct the interview at the site of the activity, while the interviewee is performing the activity.

I've found it best that the developers contact these users directly, after they have agreed to be interviewed. During a preliminary phone conversation, the developer can explain the purpose of the interview, and can describe how it will be conducted, how long it might take, and where it will happen. The developer should also get permission to record the interview at this time. Finally, it's a good idea to emphasize how important the interviewee's cooperation will be, and to thank him or her for agreeing to participate. These preliminary phone conversations have proven to be very effective in getting interviewees into the right frame of mind.

5. Set the focus. Use the same techniques described for one-on-one interviews.

6. Replan after every interview. Use the same techniques described for one-on-one interviews.

Interview Steps

It's convenient to think of each Contextual Inquiry interview as consisting of three main stages: Introduction, Observation, and Summary.

Introduction. At the beginning of the interview it's a good idea to introduce yourself and restate the purpose of your visit. Also restate the need for the interview to take place at the site of the activity.

Next, ask the interviewee to describe his or her work activity to you. Here is where you should start up your audio or video tape recorder.

You should find out what is the cycle time of the activity—is it repeated every few minutes, or is it repeated every few years (for example, the product design cycle)? Is it very much the same each time, or is it very different each time?

Ask for information about the environment or the context, including the location, the people, the culture, and the values.

This will be summary data, but it will provide a background for conducting the rest of the interview.

Observe the customer's work or use of the product. Next, ask the interviewee to begin showing you the activity you want to study.

For *short-cycle activities,* you may be able to observe the entire activity from beginning to end, possibly even a few times.

Validating our customer's assertions can be a source for surprises that may cause us to alter our focus. When customers make assertions, *ask them to show you.* Ask to see reports, examples, demonstrations, or even simulations of what the customer has told you.

This is a powerful possibility in Contextual Inquiry. There are two important reasons why we should try to see evidence of our customer's assertions.

The first reason is that we may have misunderstood the significance of what we were told. The second is that the customer may have unintentionally misled us, either by exaggeration or by use of vocabulary we are unfamiliar with. (In rare cases, the customer may even intentionally mislead us.)

For *longer-cycle activities,* such as adjudication of a law case or the design of a product, you obviously won't have time to see the entire cycle. In this case, make sure you understand the entire cycle in abstract terms (as described to you during the Introduction phase of the interview). Then, focus on that part of the cycle that is current. Try to observe as much as is practical. Starting with the specifics of what is current, it is often possible to extend one's understanding of cycles that are not currently visible.

For other parts of long cycles, ask to see examples of tangible items that result from those parts. Ask for documents, models, work in progress, demonstrations, diagrams—in short, anything you can actually view first-hand. Ask how the items were produced. Ask for an explanation of each item. Ask for examples of comparable items that may explain what stays the same and what varies from one such item to the next.

Summarize. At the end, summarize the main points of what you learned. Show the interviewee your flow diagram of the activity. At each point in your summary, ask for clarification or correction to your understanding.

Sometimes, during Contextual Inquiry interviews, the developer may see the interviewee experience difficulties that the developer can remedy. It would be disruptive of the interview process to show the interviewee your remedies during the observation phase of the interview, but it would be quite appropriate to help solve the problems during the summary phase.

Simple Prototyping
Just as there is great value for the developer in experiencing the customer's environment and activities first-hand, so also is it valuable for the customer to experience the proposed product or service first-hand. A prototype provides an opportunity for such a "sneak preview."

In many cases, developing prototypes of products or services is prohibitively expensive or time-consuming. Even when prototypes are possible, they may not be available early enough in the design cycle to have a useful impact on the design.

However, inexpensive simulations of products or services can often provide enough realism to help a customer provide useful feedback. The notion of the *paper prototype* developed along with Contextual Inquiry, and it can be quite useful.

The idea is to simulate on paper, or with inexpensive materials, some essential aspects of the product or service being planned. If such a simulation is possible, then as a customer experiences and responds to the simulation, the developer can redesign the product or service instantly—at the simulated level—and test the redesign on the customer immediately.

Common examples of paper prototypes are
  • Graphic User Interface mockups
  • Hardware mockups
  • Service scripts

Graphic User Interface (GUI) mockups, or *paper mockups,* are made by drawing a large rectangle representing a computer screen on a large piece of paper, such as a desk pad. GUIs consist of many graphical objects including menu bars, windows, icons, and buttons, each with a characteristic appearance. Each object can be simulated by a Post-it Note of appropriate size and shape, with graphics or text drawn on it to resemble its actual appearance on a computer screen.

In Diagram 16-9, the desktop, consisting of the menu bar (containing "File," "Edit," "MIDI," "Compose," and "Setup" menus), is drawn on a piece of paper with dimensions 11″by 17″. The menus are written in pencil or are each on a small Post-it Note, so they can easily be changed or rearranged. The File menu commands "Open . . .", "Close", etc., are written on a Post-it Note in pencil. Likewise for the window labeled "Rhythms" and the button palette containing the triangle and circle icons. The developer will have prepared several additional Post-it Notes, each depicting an object that could appear on the screen sometime during a user's session.

The customer is presented the configuration, and asked to "use" it to perform some task defined by the customer. The customer would "use" the prototype by describing each task. For example:
  • "I would start by selecting the File menu." (The developer would then reveal the File menu options by laying down the File menu commands on the paper prototype, beneath the "File" menu item, as shown in Diagram 16-9.)

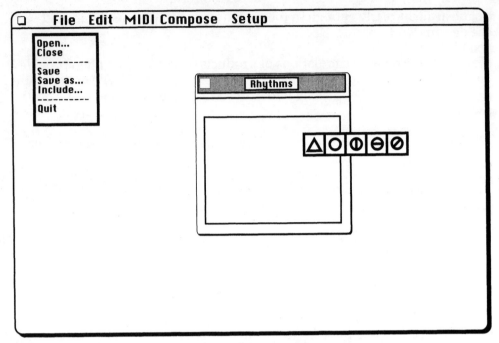

Diagram 16-9. GUI Paper Prototype

- "I want to convert a file produced by Program X, but I don't see a command here that would allow me to do that." (The developer might ask, "What command did you expect to see?", thus initiating an inquiry into why the customer wants to perform the action at this point, what controls or commands the customer expects to see, and what the customer expects the software will do next.)

  If the developer and user decide that a new command is needed in the File menu, the command can be penciled in on the spot, and the simulation can be continued.

This process could continue in as much detail and for as long as both people want it to. The entire interview would be taped or videotaped so the developer can capture all new customer needs and product concepts later on.

Hardware mockups are used similarly to GUI paper mockups, but might involve inexpensive cardboard or Styrofoam mockups instead of the two-dimensional mockups used for software simulations.

Service scripts are flow diagrams that indicate all the steps of the service *from the customer's point of view*. Once again, the developer would ask the customer to invoke the service and simulate the service by faithfully following the steps of the flow diagram. At each point, the developer would ask the typical Contextual Inquiry questions: "Is that what you expected?", "What are you trying to achieve at this point?", etc.

Analyzing Data from Contextual Inquiry

During the interview, the interviewer must be very alert to the customer, trying to understand the "meaning behind the meaning" of what the customer says. It's a good idea to take notes as the interview proceeds, however, to capture important points that you may want to come back to later. Also, experience has shown that a flow diagram of the customer's activity can be a useful way of arriving at a joint understanding of that activity. It provides an excellent way for the interviewer to quickly summarize the activity, and for the interviewee to confirm or modify that summary.

Contextual Inquiry interviews should also be recorded on videotape or audiotape. The tape is the single most important item that you will bring back from the interview. Most of the data derived from Contextual Inquiry comes from analysis of these tapes.

Initially I recommend that you transcribe the tape, word for word, onto a hard copy document, and then capture the customer data from the transcription (while simultaneously viewing the videotape if you have used video). Later on, after developing some experience in transcribing, most teams will skip the transcription step and work directly from the tapes.

Teams of two or three developers scan the transcriptions or the tapes for concrete data, and write each piece of customer data onto a Post-it Note or card. The cards can be affinitized and grouped, just like all other customer data. For more details on classifying the data, see "Analyzing Customer Data" in Section 16.5.

16.3.4   How Many Interviews?

The purpose of interviews of any type is first to identify all customer needs, and second to identify whatever innovative solutions to meeting those needs that customers may offer.

The number of interviews required is based on the yield of the interviews. The first interview will produce a great deal of new information that the interviewers will not have seen before. The next interview will usually contain some of the same information yielded by the first interview, and some new information. With each subsequent interview, we can expect to see an increasing proportion of information that we have already learned, and a decreasing proportion of new information. At some point the amount of new information from an interview will not be worth the effort to obtain it.

Since the number of interviews is going to be very small compared to the customer base (for most markets), there are no valid statistical inferences that can be made from what is learned from the interviews. Once we have heard a customer need once in any interview, we don't need to hear it again. Later, by surveying a sufficient sample of customers, we can determine how important the need is relative to all the other needs.

John Hauser and Abbie Griffin modeled the rate of new customer needs yielded by successive interviews. They concluded that twenty to thirty interviews were required to get ninety to ninety-five percent of customer needs. This yield rate is based on two important assumptions:

- Customer interviews are one-on-one.
- Each customer interview is considered to be independent of the others.

The second assumption is very important, because in practice the customer interviews *are not* independent of each other. Any interviewing team will compare notes after every few interviews, and each interviewer will remember the information educed by previous interviews. By steering each interview into new territory, the yield rate of new needs can be made much higher than the Hauser/Griffin model.

Thus, interviewers may regard the twenty to thirty interviews as the upper bound of interviews needed to get ninety to ninety-five percent of the needs.

The Hauser/Griffin research was confirmed by one-on-one conference room style interviews. No research has been done on other styles of interview, such as the Contextual Inquiry. Contextual Inquiry interviewers generally feel, however, that after about ten Contextual Inquiry interviews the number of new needs is not worth the additional effort. Considering the much higher cost of Contextual Inquiry interviews, this informal finding is somewhat comforting.

16.3.5 Focus Group or One-on-One Interview?

In many cases, the decision to use focus groups or one-on-one interviews is predetermined by external factors, such as the availability of interviewees, or the skills and preferences of the interviewers. But when all else is equal, which is the better choice? The goal is to determine as many unique customer wants and needs as possible (within a budget), so the interviewing style that is likely to deliver the most wants and needs for a given cost is the one to choose.

Hauser and Griffin have studied the relative effectiveness of focus groups versus one-on-one interviews.[5] The effectiveness of the interviews was determined by counting the number of unique customer needs generated by each interview style, and comparing them to the total number of needs generated by both interview styles.

They found in their study that four one-hour, one-on-one interviews were about as effective as two two-hour focus group interviews with six to eight interviewees present at each focus group interview.

On this topic, Griffin and Hauser conclude: "If it is less expensive to interview two consumers for an hour each than to interview six–eight customers in a central facility for two hours, then [our data suggests that] one-on-one interviews are more cost efficient."

16.4 Reactive versus Proactive Modes

Qualitative customer data is acquired in two contrasting modes: proactively and reactively.

The proactive mode involves intentional steps by the development team to seek information. In this mode the development team takes control over the amount and type of information they will receive, as well as the timing. Surveys, focus groups, and one-on-one interviews are examples of proactive methods of listening to the customer.

The reactive mode involves customer data arriving at the doorstep of the development team. In this mode the customer takes control over the amount, type, and timing of information transmitted. Customer complaints, litigation, and letters of appreciation are examples of sources of reactive customer data.

Reactive data is generally more plentiful and more available to developers than proactive data. Customer complaint databases are generally developed in order to help organizations defend themselves against complaints that arrive faster than the organization can respond to them. Backlogs of complaints often build up. They must be organized to enable specially formed teams to handle them.

Most companies do a good job of organizing and keeping track of incoming negative customer data. Positive data may find its way into sales organizations in order to help with the selling effort, and it may be used to congratulate employees who have contributed somehow to a success story. However, because positive data rarely needs follow-up, it usually doesn't get the same kind of management attention that is given to negative data.

As a result, reactive data is generally most useful for discovering what makes customers unhappy. In other words, it provides clues to customer needs that would classified as "Expected" in the Klein model. Because customers don't separate needs from technical solutions, these complaint databases usually also contain rich information about Dissatisfiers (from the Kano model).

Customer complaint databases are difficult to analyze from a QFD perspective because of the way they are typically organized. Most complaint databases classify complaints by urgency or priority of response needed. As a rule, the urgency of the complaint is not even assigned by the customer, but by the complaint handling team.

High-priority complaints are those that directly or indirectly will cause a lot of trouble to the complaint management team if not responded to quickly. High-priority complaints are usually associated with failures of a product or service that render it completely useless to the customer, or that create a safety hazard or other catastrophic impact on the customer.

Lower priority complaints are generally related to the other categories of customer need (Hidden, High Impact and Low Impact). Because they are lower priority, however, they sometimes receive little attention, and particularly, little description.

The potential for learning a lot about Expected needs and Dissatisfiers from customer complaint databases is high, because there is usually a lot of data. Because the databases are not organized to support QFD, however, they require much resource-intensive analysis before they can provide their potential benefit.

Development teams should examine complaint databases and other forms of reactive customer data to determine the cost-effectiveness of gaining useful information for QFD. Random sampling of these databases could be a way of harvesting useful information at a reasonable cost.

More important, development teams would do well to enhance the structure of these databases for future incoming complaints, so that the incoming data will be easy to access in the future.

16.5   Analyzing Customer Data

After transcripts of interviews or focus group sessions of any type have been prepared, the development team is faced with the task of making sense of the data. In particular, they must create an affinity diagram of customer needs in order to begin constructing the House of Quality.

The overall process to produce the affinity diagram is this:

1. Identify phrases that represent customer needs and copy them to cards or Post-it Notes. Wherever possible, use the actual words of the customer. Try to use statements about concrete experiences, rather than statements that summarize feelings. Later, in the affinity diagramming process, a collection of concrete statements will be generalized by the development team into a higher level customer attribute.
2. Sort the phrases into true customer needs and other types of information, using a construct such as the Voice of the Customer Table (Section 5.2.) Along with the sorting process, development team members will undoubtedly develop questions, issues to be resolved, and product concept ideas. These should be sorted into appropriate categories, to be used or otherwise dealt with as the QFD process moves along.
3. For the remaining customer needs, create an affinity diagram.
4. Select the secondary or tertiary level to represent the Customer Wants and Needs in the House of Quality.
5. Document all results, especially any unexpected results. Be prepared to inform the organization about these results.

16.5.1 Who Creates the Affinity Diagram?

The expert on how customer phrases go together, and what they mean when they are grouped, is the customer. With the exception of the VOCALYST[6] process, which will be described below, the affinity diagramming process can usually be influenced by a few customers at most.

If VOCALYST is not available to the team, a reasonable substitute is to invite a few key customers to the affinity diagramming process, and either ask them to build the diagram, or to participate with the developers in building the diagram.

If customers are not available, the development team will have to build the affinity diagram by themselves. The team must be careful to distinguish between customer needs and technical solutions, and to think like a customer as much as possible. There is an obvious risk involved when customers do not participate in this part of the process.

16.5.2 VOCALYST

VOCALYST is a proprietary Voice of the Customer process. It is unique in that it uses statistical methods to create a tree diagram of customer needs representative of how customers have actually affinitized those needs.

VOCALYST begins with a series of one-on-one interviews of customers or users of competitive products or services. These interviews are conducted in the "probing" style that we have referred to earlier.

Each interview is transcribed word for word. Each transcript is analyzed by at least two experienced readers, who independently highlight every phrase that represents a customer want or need. All these phrases are entered into a computer database. Typically, twenty transcripts will yield about 1000 customer phrases. Using proprietary methods, these phrases, which include many duplicates or similarly worded needs, are reduced down to a set of 50 to 150 unique wants and needs.

Each need is noted on a card, and a full set of cards (usually between 50 and 150) is sent to approximately 100 customers. They are requested to group the cards in piles that seem to them to be similar. In essence, each customer is asked to create a single-level affinity diagram. This single level will eventually become the tertiary level of the final customer need hierarchy.

The customers then pick the need (card) in each pile that seems to them to best represent that pile, and put it on top of the pile. The cards placed on top of the piles are called "exemplars" and are used in later processing to help the VOCALYST analysts determine the names or titles of the secondary and primary levels of the customer need hierarchy.

Next, the customer picks the pile that is most important to him or her, and marks it 100. All of the other piles are then marked with a number that represents its importance relative to the most important pile (the one marked 100). This step captures weighted importance of the groups of customer needs.

The customer is also requested to evaluate how well their current product meets the need represented by each pile. This is the customer satisfaction performance value for the customer-defined secondary need. The customer then returns the cards to be analyzed. Typically the return rate is about seventy percent.

Each respondent will sort the deck somewhat uniquely. The VOCALYST process uses proprietary statistical techniques to develop a composite view of the secondary attributes across all respondents, as well as the importance and customer satisfaction performance levels for each secondary attribute. This is exactly what is needed to begin constructing the HOQ.

16.6   Quantifying the Data

Once the team has its affinity diagram, they are ready to quantify the data. The data required for QFD is

- Relative importance of needs
- Customer's satisfaction performance level for each need
- Competition's customer's satisfaction performance level for each need

Most commonly this data is collected by survey. We've discussed these quantitative methods in Chapter 6.

The key pitfalls of constructing surveys are

- Selecting appropriate samples and sample sizes
- Ensuring adequate response
- Wording the survey questions to avoid ambiguity
- Analyzing the results

Although surveys are most efficiently conducted by professional market research organizations, many development teams conduct their own surveys. Most "home-grown" surveys take much longer than planned, and thereby engender hidden costs.

Sometimes development teams judge that they cannot afford the time or the money to conduct a survey. This is generally a short-sighted judgment, driven by poor budgeting or cost-accounting. Nevertheless, such teams can and do assign the quantitative data by team consensus during the QFD process.

16.7   Voice of the Customer Overview

Diagram 16-10 illustrates the various techniques for acquiring the Voice of the Customer and for getting it ready for QFD. The first step is to talk to customers. There are three primary methods for doing this.

The focus group is generally quick and inexpensive compared to the other methods, but the customer needs yield is the lowest of the three main interviewing methods.

Contextual Inquiries are time-consuming and costly, because members of the development team must travel to the customer's site for each interview. The customer need yield is very high.

Conference room interviews provide a middle ground in terms of time, cost, and yield of customer needs. I recommend a blend of some initial Contextual Inquiries, followed by conference room interviews.

Once the interviews have been conducted, the development team will analyze the transcripts of the interviews and extract the customer phrases. There are likely to be hundreds or even thousands of phrases. Many of which will be duplicates and many of which will be technical requirements, Substitute Quality Characteristics, and other phrases that are not actually customer

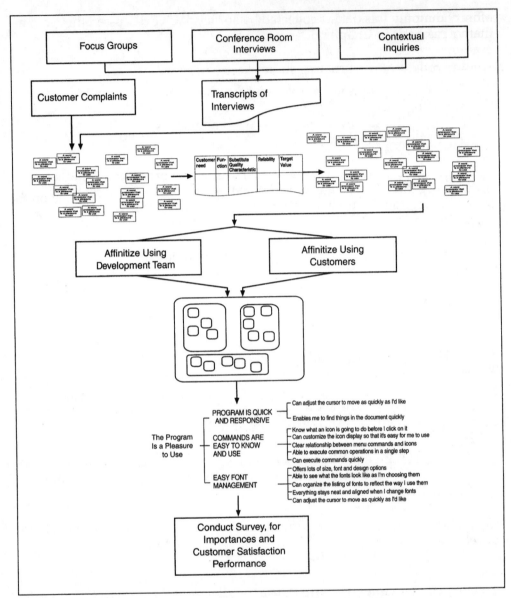

Diagram 16-10. Voice of the Customer Overview

needs. The team will have to sort these phrases, using a mechanism like the Voice of the Customer Table (Chapter 5). All the phrases are important, but only the customer needs phrases are needed at this stage.

The customer needs phrases can be transcribed onto cards or Post-It Notes and structured using the affinity diagram process (Chapter 3). Ideally, the affinity diagram would be strongly influenced by the customer, using a method like VOCALYST, or by inviting customers to assist in the affinity diagramming process. Lacking customer involvement, the development team would create the affinity diagram.

Next, the team will restructure the affinity diagram into a tree diagram. This is partly a clerical process of redrawing the diagram. However, it also involves ensuring that the structure is complete at all levels, because it's possible that a few customer needs will have been missed in the interviewing process.

The team next decides which level of the tree diagram will be used in the House of Quality. Generally the secondary level is used, but depending on the size of the tree diagram, and the problem the team is trying to solve, other levels, or even subsets of a certain level, may be used. For example, if the project is to improve the packaging of the product, only those customer needs pertaining to packaging would be used in the QFD.

The final step is to determine the importance and customer satisfaction performance of the needs to be used for QFD. This is normally done by survey. Whereas the interview stage necessarily involves no more than fifty, and often fewer, customers, the survey would reach hundreds or even thousands of customers in order to assess the majority of customers, and to understand the way they are segmented.

The Center for Quality Management (CQM) is a Massachusetts-based consortium of companies that share the best Total Quality Management practices. The product development practices they espouse have been strongly influenced by their membership, and by Professor Shoji Shiba.[7] Their product development process model contains a step that they describe as "making customer requirements visible and explicit." This step contains three substeps, which they call stages.

Stage 1 calls for understanding the customer's needs and environment, which can be accomplished by interviews and especially by Contextual Inquiry.

Stage 2 calls for converting the understanding of the customer's needs into product requirements. This involves first identifying those statements from the interviews that represent customer needs, and then translating them into solution-free statements of what the product should do. This is similar to the work required in filling out the Voice of the Customer Table. The team then selects the most important or significant customer needs. This is done by team judgment based on the team's understanding of the customer and what the team heard from a limited number of customer interviews. A more rigorous approach would involve a survey process, as described in this chapter, and also shown in Diagram 16-10.

In Stage 3, the team develops a rank-ordering of the remaining customer needs. One approach is to classify the needs by the Kano model. The team then generates metrics for the customer requirements. In our language, the team generates performance measures for their SQCs. They then create a QFD matrix (which Shiba *et al.* call a quality table), in which the customer's needs are the "Whats" and the performance measures are the "Hows."

Thus, this "CQM" variation on acquiring the Voice of the Customer uses techniques similar to those described here. Many variations on the methods described in this book are possible. The best strategy is to understand the many techniques and methods available, and to choose the ones that best meet the needs of each development team.

Summary

Gathering the Voice of the Customer for QFD consists of two distinct activities: first, gathering qualitative data; second, quantifying the data.

Useful models exist for classifying customer needs. One such, the Klein model, provides a structure for customer needs that mirrors the Kano model for technical characteristics.

There are many methods available for gathering qualitative data: reactive and proactive methods, and interviews of many types. Developers will understand their customers' needs best if they are in direct contact with them. Recording and transcribing interviews with customers frees the interviewer to concentrate on the job of interviewing. It also provides a large volume of unfiltered data for subsequent analysis.

Analysis of qualitative data involves extracting the Voice of the Customer directly from the transcripts of interviews, and using these phrases to construct an affinity diagram. Customer input in the affinity diagram process provides the best assurance that the resulting structure truly reflects the way customers think.

Once the affinity diagram has been constructed, the data is ready for quantification. The most reliable method for quantifying is a well-designed survey responded to by an adequate number of customers. If a survey is not possible, many development teams estimate these numbers themselves.

Once the Voice of the Customer has been gathered, sorted into a hierarchy, and quantified, the development team is ready to build the House of Quality. We've already had a good look at the structure of the HOQ in Part II. The next chapter concentrates on QFD implementation topics not yet discussed.

Discussion Questions

Compare the methods your organization uses for gathering and analyzing the Voice of the Customer to the methods outlined in this chapter. What are the advantages and disadvantages of these methods for your organization? Taking all factors into consideration, if you were to launch a Voice of the Customer project right now, which method would you use? What would it cost? How long would it take?

[1] Other than presentation notes distributed at a meeting of the QFD Institute (Ann Arbor, Michigan), the present text provides the only published description of this model.

[2] The segment/users grid in Diagram 16-7 resembles a three-factor "Designed Experiment." Taguchi's orthogonal arrays, or other fractional matrices, could be used to plan out groups to be interviewed.

[3] M. Good, "The Iterative Design of a New Text Editor," Proceedings of the Human Factors Society, 29th Annual Meeting (Baltimore, 1985), pp. 571–574.

[4] D. Wixon, K. Holtzblatt, and S. Knox, "Contextual Design: An Emergent View of System Design," Proceedings of CHI'90, Human Factors in Computing Systems (Seattle, April 3–5, 1990), ACM, New York.

[5] Abbie Griffin and John Hauser, "The Voice of the Customer," Report Number 92–106, Marketing Science Institute, Cambridge, Massachusetts, March 1992.

[6] Robert Klein, "New Techniques for Listening to the Voice of the Customer," Applied Marketing Science, Inc., Second Symposium on Quality Function Deployment, June 18–19, 1990, p. 197.

[7] Shoji Shiba, Alan Graham, and David Walden, "A New American TQM: Four Practical Revolutions in Management," Productivity Press/Center for Quality Management, August 1993.

CHAPTER 17

Phase 2 and Phase 3: Building the House and Analysis

Introduction

This chapter will discuss the procedure for constructing the House of Quality. It will focus on the group processes, sequencing of events, and the logistics of the team working together. It will focus on the practical considerations that come into play. It will deal with consensus processes and practical issues such as how long a meeting should last, and how it should be run and organized.

Building the House of Quality is often thought of as "the QFD process." We have seen that the QFD process encompasses much more than assembling a team in a room and constructing a large matrix. A successful QFD process depends on good planning, good customer data, the right team, and good Substitute Quality Characteristics, all of which should be collected or prepared for in advance. However, making good use of them during the House of Quality construction meetings is also critical to the success of the QFD process.

17.1 Sequencing of Events

The most useful sequence for building the House of Quality (Diagram 17-1) is this:

- Construct customer needs/benefits
- Construct planning matrix and analyze results thus far
- Generate Substitute Quality Characteristics and analyze results thus far
- Determine Relationships and analyze results thus far
- Determine Technical Correlations and analyze results thus far
- Acquire competitive benchmarks and analyze results thus far
- Set Targets and analyze results thus far
- Plan the development project, based on the results

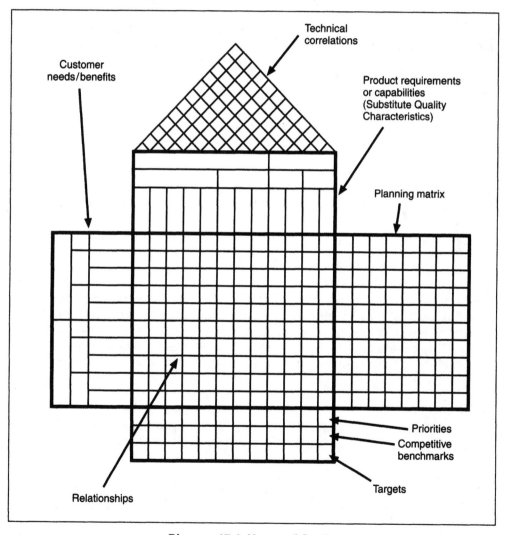

Diagram 17-1. House of Quality

Notice that analyzing results (Phase 3) at every step is woven into the process of building the HOQ (Phase 2). In this book, we have identified analysis as a separate phase of QFD in order to highlight the activity. In practice, explicit time slots for analysis should be allocated, but throughout the QFD process (at the end of each of the sequenced steps), rather than only once, at the end of the entire process.

Construct customer needs/benefits. These are normally acquired via market research. Hence, they are normally known when construction of the House of Quality begins.

If customer needs have been acquired through any method or combination of methods described in Chapter 16, then the team must be familiar with these needs before the HOQ construction begins. Alternatively, the first step in construction must be a familiarization process. The more time the team spends on getting familiar with customer needs, the better.

At a minimum, the QFD team members must be familiar with the structure of the customer needs—the primary, secondary, and tertiary attributes. They must also understand the relative importance of the attributes at the level being used for the QFD; and they must understand satisfaction performance levels. This basic data will be referred to continually throughout the QFD process.

Construct planning matrix and analyze results thus far. Customer importance and satisfaction performance data normally comes from market research and can simply be transcribed into the matrix. The remaining parts of the planning matrix, setting goals and determining sales points, are strategic planning activities and should occur before the more detailed aspects of planning proceed. The end result of the planning matrix is the calculation of raw weights of the customer needs. Very often, especially when time is limited, analysis of the rest of the HOQ proceeds with the most important of these customer needs and may exclude the least important. This selectivity cannot occur unless the planning matrix is complete.

After the raw weights have been calculated, most teams devote some time to analysis of the results thus far. The raw weights (or normalized raw weights) of the customer attributes are often displayed in a bar chart, as in Diagram 17-2.

The bar chart in Diagram 17-2 bears some resemblance to a Pareto diagram, because it graphically displays numeric information in descending sequence, and because it helps the team to focus on the most important attributes.

It is not, however, a Pareto diagram. Pareto diagrams are typically created by counting defects that have been classified by category. In contrast, the

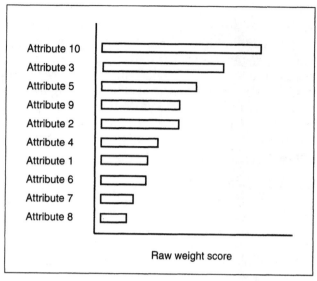

Diagram 17-2. Bar Chart of Raw Weights

attribute bar chart of raw weights displays the results of market data and team judgments converted to numbers and multiplied together. Another distinction is that Pareto diagrams typically reveal that the one or two most important items far outweigh all the others (an evocation of the "80–20 rule"). In QFD it is rare for the most important items to vastly outweigh the others: the decline is generally more gradual.

Analysis usually begins with an assessment of expected and unexpected results. The team will trace the steps it took to arrive at the resulting raw weights, and provide explanations for why the raw weights turned out as they did. They may go back and make adjustments, based on the learning that took place since the start of the QFD process. They may do some "what-if" analysis, to see whether some of their decisions may have had far-reaching implications upon the results (in which case they may revisit these decisions to be sure they have come to the right conclusion).

The analysis process is very important for team building, as well as for the team to develop confidence in the QFD process. The team must be comfortable with the rank-ordering of the customer attributes before they can proceed.

Generate Substitute Quality Characteristics and analyze results thus far. The team uses any of the methods described earlier to generate the SQCs. As has been described in Chapter 7, this is generally the most challenging and time-consuming part of the HOQ process.

Once the SQCs have been generated, the team should allocate one to four hours to consider the process they just went through. In most cases, generating the first SQC takes a long time, as much as half a day. By the end of the process, most teams can generate SQCs almost as quickly as the facilitator can announce the customer attributes that initiate the discussion. Obviously some sort of learning process has occurred. The learning inevitably must have resulted in some new shared perspectives on the product planning process. This is the right time to make this learning process explicit, by conducting a formal "analysis" session. As with all of the Phase 3 sessions, analysis should consist of the following:

- Retrace the steps taken so far.
- Consider adjustments to earlier parts of the process, based on what the team has learned since those parts were completed.
- Review and document all assumptions made.
- Review and document unresolved issues.
- Develop action items to resolve all unresolved issues.
- Develop a list of next steps.
- List the benefits the team has derived from the process thus far. The facilitator's job at this point is to focus the team on what they must do after the QFD process has finished, in order to move the development project along. Some actions will involve preparing for the remaining QFD steps, such as completing the roof. Other actions will directly relate to the development work. Still other actions will relate to informing the rest of their organization about their decisions and findings.

Determine Relationships and analyze results thus far. This section is the natural next step after generating the SQCs. No other part of the HOQ can be completed sensibly until the Relationships and resulting priorities have been determined. Larger teams (more than ten people) generally break into subteams to complete this step.

When the Relationships have been completed, the team then calculates the priorities of the SQCs (generally by means of QFD software) and performs an analysis step similar to the ones previously described.

Once again, analysis at this point is a critical step that allows the team to understand the implications, benefits, and shortcomings of the process they have just gone through. What-if analysis on controversial decisions will help the team feel comfortable with those decisions, or will persuade them to do further fact-finding or other research in order to come to a better decision.

Technical Correlations. This QFD step is often omitted, although the author believes that if the SQCs are chosen properly, analysis of the technical correlations can yield high benefits. When it is performed, teams usually start with the highest priority SQCs. They may limit their analysis to the top ten or the top one-third of the SQCs. If they do the entire roof, they may break into subteams, although it is difficult to break this matrix into submatrices.

An analysis at the end of this step is extremely beneficial. This step will likely result in many important actions items that affect communication among organizations during the development process.

Acquire competitive benchmarks and analyze results thus far. Measuring the competition's critical SQCs cannot be done in a conference room. This activity is usually planned after the SQC priorities have been computed, and is executed as a project "off-line."

Depending on how quickly results come in, one or several analysis sessions will be useful. If the results are coming in slowly, say over several weeks, interim analysis sessions can help the team see how to speed up the process, or how to perform dependent steps sooner.

Set Targets and analyze results thus far. After any and all competitive bench-marking has occurred, the team normally convenes to set Targets. An alternate approach is for a committee to propose Targets and present them to the rest of the team for approval.

At this point, the HOQ is complete. The team must not omit a final analysis session to consolidate all gains, to ensure that the development process builds on the QFD results, and to make recommendations for more effective QFD processes in the future.

17.2   Group Processes/Consensus Processes

The QFD manager's role was discussed in Chapter 14. It's helpful to define other formal roles as well. These are the QFD facilitator, the recorder, and the HOQ scribe. Sometimes more than one of these roles can be taken by a single person. It's worthwhile, however, to define them separately in order to best understand what they are.

17.2.1 The QFD Facilitator's Role

Drive the planning process. Make sure that the elements of planning, as described in Chapter 15, get done. Many facilitators develop a checklist of planning questions and use it to ensure the planning process is completed.

Allocate time. Set up a detailed schedule, using the guidelines in the QFD Estimator Chart. Develop an agenda for each group meeting. In the agenda, show the objectives for each meeting, and show how these objectives relate to the overall QFD activity. Managing time will be much easier as a result of this type of preparation.

Explain each QFD step. Even though the team may have received some training in QFD, it's a good idea to explain what will happen in the step that's about to begin. In my experience, many teams don't really understand the process until they have experienced it; hence, they may flounder at the beginning of a step. Often it's helpful to practice the step on something simple and unrelated to the development project—although many concrete-minded engineers prefer practicing on the project itself.

Get consensus. The QFD process may be thought of as an almost endless succession of consensus events. At every small step of the way, the team must agree on something. Each cell of the HOQ matrix represents a question, and the team must agree on each answer. In the interests of time management, the facilitator must have a supply of tools aimed at helping the team reach consensus literally thousands of times, quickly.

Here are some general principles for getting rapid consensus:

1. **Ensure that the group knows exactly what the consensus topic is.** The simplest technique for the facilitator is to state the question the team must agree on, and provide a format for the answer. For example,

   "If we could move SQC_X, how much would be the change in satisfaction performance on Customer $Need_A$: big change, medium change, small or no change, or definitely no change?"

2. **Avoid discussion that doesn't contribute to the team's understanding of the topic.** Often team members will appear to be arguing, but a careful facilitator may notice that they are actually in agreement (sometimes called "violent agreement"). They may not realize it, because they may be using different language to say the same thing. This is common when the team members come from different disciplines, as is the case in cross-functional teams. The alert facilitator will be on the watch for violent agreement and help the team to see it when it occurs.

   A good technique for preventing violent or time-consuming agreement is to use silent techniques to quickly sense what everyone is thinking. The silent vote works well:

   First, ask the question: "If we could move SQC_X, how much . . . ?" Then, ask each team member to silently consider the answer, and silently raise a number of fingers indicating his or her choice (as in Diagram 17-3.)

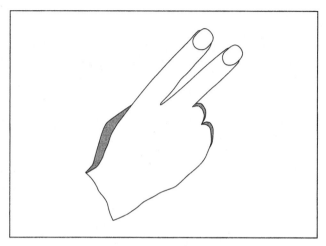

Diagram 17-3. Silent Vote

The facilitator can quickly assess, by a show of silent votes, whether everyone is in agreement (no discussion necessary), whether just one or two people are voting slightly differently than the majority (perhaps they could "live with" the majority decision), or whether there is significant polarization.

If there is polarization, it is usually most fruitful to ask a person voting in the minority to present an explanation for his or her vote. Very often, the minority view turns out to have been overlooked by the others. Once they hear the minority view, many people may quickly switch. If the group had simply started by discussing the majority view, almost no forward progress would be made, since the majority of the group already holds that view—hence the emphasis on the minority view.

As soon as the minority has had a chance to explain their position, it's a good idea to get the group to vote silently again. Some people may have changed their minds, and others may be able to "live with" the new majority.

3. **Create a sense of urgency.** The enemy of any meeting is time. A good facilitator will set interim goals for the team, such as "Let's complete fifty cells in the next two hours", and then continually remind the team of how much time is left. This frequent reminder will help each team member to use the team's time with care.

4. **Clarify for the team any differences of opinion.** When the minority *can't* live with the majority, the team needs help to move ahead. One very useful service the facilitator can perform is to understand what the legiti-

mate difference of opinion is, and feed it back to the team. Very often, this will help someone on the team to see a way of reconciling the two sides. If not, the conflict can be documented as an unresolved issue. Since the facilitator will see to it that all unresolved issues get converted into actions to resolve the issues, the team will have confidence that the problem will eventually be straightened out, and they can move on.

5. **Check frequently for shifts of opinion.** Individuals often change their minds silently and the change must be explicitly revealed. This has already been covered in point 2.

6. **Seek realistic consensus.** "Realistic" means that most team members agree, and the remaining members can "live with" the decision. Realistic consensus, in which team members are generally aligned if not precisely so, is usually good enough for development efforts. Only in a disciplined process such as QFD can such detailed agreements and disagreements be documented and worked through. Before the advent of QFD, development teams did their work in the context of much more severe, and much more hidden, disagreements.

Wherever a team is concerned about their decision, they can make a note of their concern, and perform "what-if" analysis on the completed matrix. The nature of "what-if" analysis is: set the value of the cell in question to one extreme value, compute the resulting priorities (or raw weights), then set the value to the opposite extreme value and recompute the priorities. If the priorities or raw weights shift by only a little, then that decision can be seen to influence the project only by a little. If the priorities shift by a lot, then the team knows it must revisit its decision to be sure they can believe in and support its implications.

17.2.2   The Recorder's Role

The recorder's job is to record all assumptions, decisions, and action items that arise from the meeting. Once recorded, they should be distributed to the team as quickly as possible after each meeting, since many of the actions may be due for completion before the next meeting starts. Experience shows that the facilitator usually cannot concentrate on the recorder's duties, and therefore needs another person for this task.

Most commonly, development teams rotate the responsibility throughout the group.

17.2.3   The HOQ Transcriber's Role

Determine the mechanism for creating the physical charts. The task of creating a QFD chart that covers the wall, complete with neat rows, columns, and labels large enough for team members to read at a distance, is mundane yet critical. It is so mundane it can be easily overlooked. Nothing is as embarrassing to a facilitator, or as boring to a team member, as creating the chart at the last minute, while the team watches with nothing useful to do.

Likewise, the calculations are simple enough, but time-consuming and error-prone. No team that has sat with hand calculators performing hundreds or thousands of simple calculations will be motivated for QFD in the future.

These mundane tasks must be planned for, and they must proceed effortlessly during the QFD meetings. In the mid- and late eighties, when little or no QFD software existed, and when laptop computers were also rare, many teams were forced to manually create the QFD charts and manually calculate raw weights and technical priorities. There is simply no reason any longer to do so. Several QFD software packages exist, and they run on the popular operating system platforms. Most of these packages will print large, empty QFD charts on oversized paper. Even empty charts printed on multiple $8\frac{1}{2}''$ by $11''$ sheets can be enlarged on special copying machines and pieced together into a large chart. These charts can then be taped to the wall and filled in by the facilitator as the QFD process moves on.

A word to the wise: Don't let this seemingly simple task ruin a QFD activity. Work out the details in advance.

Know the QFD software. Being familiar with any tool helps the tool wielder to do a better job. QFD software does not require software hackers to operate it, but the options are generally rich enough to require some preparation and learning before using it in a QFD meeting. Once again, be prepared.

Acquire all the physical support. The physical support for QFD has been identified and discussed in Section 15.8. As with the QFD charts and software, be prepared.

Publish intermediate versions of the QFD charts. When the QFD process is going smoothly, the team is adding information to the QFD chart continually. Most team members appreciate having copies of the chart in hand, as up-to-date as possible. Even with large charts on the wall, many team members prefer to look at a chart right in front of them. (The chart on the wall still serves the purpose of focusing the team on a single objective.) Keep the team up-to-date with frequent printouts of the matrix or matrices.

Summary

In this chapter we have seen how to sequence the building of the House of Quality. We have discussed the benefits of reflecting on the QFD process as it progresses.

These benefits include understanding explicitly the implicit decisions and learnings of the team, gaining confidence in QFD and in the team's ability to use QFD for their development work, and helping the team focus on the work they must do to capitalize on the QFD results. This reflection process is important enough to call out as a separate phase of QFD, even though it occurs at the end of each QFD step, as well as at the end of the entire process.

We have seen that the sequencing of QFD steps is crucial. The facilitator should carefully map out the sequence in advance, using the recommended sequence in this chapter as a guide.

Many mundane aspects of running the QFD process can create havoc if not worked out in advance. These aspects include preparation of the matrix as a wall chart, the use of QFD software, and the acquisition of the proper room and materials.

We have pointed out that chances for success in QFD can be enhanced by ensuring that certain roles are filled: especially the QFD manager, the QFD facilitator, the recorder, and the HOQ transcriber.

If you have come this far, congratulations! Completing a House of Quality is no easy task, but the effort is well worth it. Now that the HOQ is finished, it's time to consider how QFD can continue to help the development team. In Part V, we'll look at QFD beyond the House of Quality. We can think of Part V as describing "advanced" QFD, a topic for organizations that have achieved substantial maturity in the Quality domain. Let's see what advanced QFD looks like.

Discussion Questions

Based on your own experience, list the principles of good meeting management. Map these to the techniques for planning and running QFD. How closely do they match up? Where are they different? What accounts for the differences?

PART V

Beyond the House of Quality

CHAPTER 18

Beyond the House of Quality

Introduction

Most organizations that use QFD stop after developing their customized version of the House of Quality. In some cases, some groups extend their analysis to an additional matrix in which performance measures from the matrix are deployed against features of a product or service. In a very few cases, organizations have gone further, even as far as matrices or tables that describe shop floor processes and machine settings. Even in Japan, where QFD originated, the majority of QFD applications stop with the HOQ.

There are numerous reasons why QFD teams don't use the full possibilities of QFD.[1] Partly this is due to lack of specificity in the literature as to how to use downstream QFD matrices. The lack of specificity is an inherent problem in explaining QFD. Real case studies are hard to find—companies are reluctant to share them with the public. Artificial case studies are not very convincing, and in any case the more specific the example, the less likely it is to resemble any particular reader's experience.

This chapter will describe the prevailing QFD models. It will present some general principles that should help the QFD practitioner feel comfortable using these models or modifying the models for particular applications.

There are two dominant QFD models in the U.S.: the "Four-Phase Model" and the "Matrix of Matrices." While at first glance one may feel required to choose one or the other, in fact they are not in conflict.

There are several important reasons why they are not in conflict.

The first reason is that neither model was ever intended to be used as presented. Any presentation of QFD or any other complex idea has to be specific enough for the audience to understand it. Unfortunately, the specificity carries with it the suggestion that the audience must accept it all or reject it all. In fact, the intent behind both models is to present a basic structure to be customized for each application. So what we really have are two models, each of which can be modified by adding or subtracting matrices, or by redefining matrices presented in the model. It may be stretching things only a bit too far to say that a team could start with one model, and add and subtract some matrices, only to find themselves implementing the other model.

The second reason is that the smaller, Four-Phase Model is contained within the Matrix of Matrices. Thus, if you implement the Four-Phase Model, you have implemented a subset of the Matrix of Matrices, and if you implemented the Matrix of Matrices, you will have implemented all of the Four-Phase model.

This second reason leads us to an understanding of the differences between the two models.

The Four-Phase Model is a blueprint for product development in a mature, efficient, disciplined organization. The Matrix of Matrices is a blueprint for product development in such an organization, but also within the environment of Total Quality Management. The Four-Phase Model covers basic product development steps. The larger model explicitly covers a host of activities that are not explicit in the Four-Phase Model, although at least some of them are probably performed during product development anyway. These additional activities include such topics as reliability planning, manufacturing quality control, value engineering, and cost analysis. A standard text on product development such as Clausing's[2] covers all these activities, but they are positioned as adjuncts to QFD rather than being a part of it.

Thus, the differences between the two models can be seen as matters more of style than of content. Serious students of QFD will be familiar with both. They will draw from the simpler or the more elaborate model, depending on the receptivity of their audience.

Now let's look at these two models in more detail.

18.1   The Clausing "Four-Phase Model"

Probably the most widely described and used model in the United States is a four-level model known as the Clausing model,[3] or the ASI model. ASI is the American Supplier Institute, an organization that has done much to popularize QFD. Diagram 18-1 depicts the model.

The House of Quality is similar to the one described in this book. The SQCs chosen for the HOQ are system performance measures. The prioritized performance measures are transferred to the left of the second matrix.

The process for arriving at the part characteristics in the Design Deployment Phase is not obvious from the diagram. Part characteristics are obtained through the process depicted in Diagram 18-2.

Design Deployment (Part Deployment). The first step in part characteristics deployment is to develop a function tree as described in Chapter 7. In Diagram 18-2 we see the total product first broken into subsystems, and the subsystems broken into parts. At this point, the important characteristics of each part are enumerated (the part characteristics). These will be descriptions of the parts that are critical to their design. They may include measurements, with directions of goodness, which in effect are specification parameters for the parts.

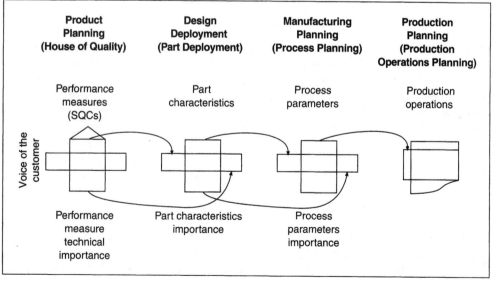

Diagram 18-1. Four-Phase QFD Model (Matrix Titles in Parentheses Are Clausing's)

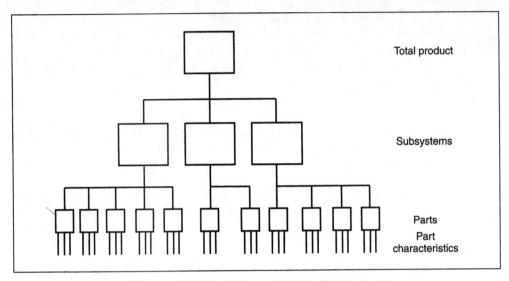

Diagram 18-2. Part Characteristics Deployment

In Diagram 18-3, the software package is the total system. Its subsystems include the contents of the carton that the software is to be shipped in, as well as the carton itself. In this case, the development has decided that the carton should serve as permanent storage for the manuals. Thus, "Provides Permanent Storage for Manuals" becomes a part characteristic. Other part characteristics, such as "Fits on Retail Store Shelf" and "Protects Contents During Shipping," are stated as phrases and also as measurable part characteristics.

The function tree approach is one way of arriving at part characteristics. Depending on the dominant technology and design culture in various organizations, other methods of parts deployment and part characteristics generation would be used. In software development, for example, the function tree approach is feasible with two changes. First, "Parts" would be replaced by "modules" or "objects," which are terms that have specific technical meaning in software development environments. Second, part characteristics, which are measurable, would be replaced either by module functions, which are verbal descriptions of work that the modules would perform, or by descriptions of data structures and data transformations.

The part characteristics are placed on the top of the Design Deployment matrix. The team then estimates the impact of each part characteristic on the performance measures. The priorities of the performance measures are multiplied by the impacts to compute the relationships, just as raw weights of customer attributes are multiplied by impacts of SQCs in the House of Quality. The relationships are summed, and the resulting importance values prioritize the

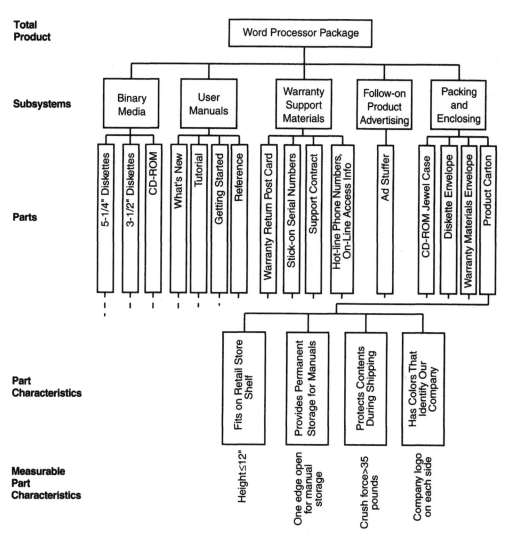

Diagram 18-3. Software Packaging Part Characteristics Deployment

part characteristics. This information tells the developers which part characteristics and which parts will be the drivers of customer satisfaction.

For complex products this process is repeated through as many levels as needed to fully specify each subsystem and part.

Manufacturing Planning (Process Planning). The procedure for manufacturing planning is also not explicit in Diagram 18-1. A recommended method for

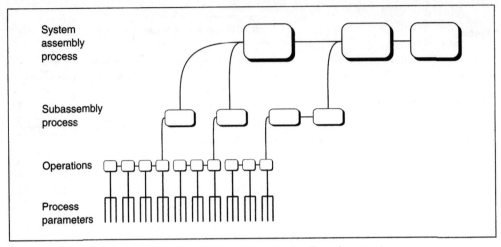

Diagram 18-4. Process Parameter Development

manufacturing planning that would result in process parameters is depicted in Diagram 18-4.

The team lays out the main process flow, or system assembly process, and then decides on the subassembly processes needed to feed into the main flow. The operations required to produce each subassembly are next added to the diagram.

This has been described as a top-down process, much like the tree-diagram process. In fact, any combination of top-down or bottom-up processes, including simply using existing process layouts, may be appropriate.

Once the operation steps have been identified, the development team uses their expert knowledge, along with experimentation, to identify the key operations process parameters related to performing the subassemblies. These parameters are specific to the operations, not the products produced by the operations, so they will likely relate to such measurements as machine adjustments.

The process parameters now become the "Hows" at the top of the Process Planning matrix. They are prioritized based on their impacts on the part characteristics.

Production Planning (Production Operations Planning). This chart is not a matrix, but rather a table or list that constitutes a checklist of topics or issues that should be considered in planning production steps. Steps suggested by Clausing and Krinninger[4] include

- Machine settings
- Control methods
- Sampling size and frequency
- Control documents
- Operator training
- Preventive maintenance tasks

These or similar topics are arranged along the top of the table, and the most important process parameters are arranged along the side. The production planner fills in the table with comments, target values or any other appropriate language. In this way, production planning can be linked all the way back to the Voice of the Customer, three QFD levels distant.

18.2   The Akao "Matrix of Matrices"

The Akao QFD model is gigantic and far-reaching. As indicated earlier, the Matrix of Matrices explicitly refers to product development steps that are not shown in the Four-Phase Model.

In the most accessible English-language source,[5] the QFD structure is presented as a system of thirty matrices, charts, tables, or other diagrams. Some are the familiar QFD-style prioritization matrices. Others are tables similar to the Voice of the Customer Table or the Production Operations Table. The thirty matrices are generally presented as a grid of matrices. The first four rows are numbered 1 through 4, and the columns are labeled A through F. A fifth column, labeled G, contains six matrices labeled G1 through G6. Thus, any matrix can be referred to by its coordinates—for example, "B3." The layout of the matrices is detailed in Diagram 18-5.

Matrix A1 is the familiar House of Quality (except for the roof, which is Matrix A3). Some other matrices correspond to the tables in the Four-Phase QFD Model. Depending on the selection of Whats and Hows in each matrix, additional matrices are used to analyze such topics as

- Competitive analysis versus cost
- Parts versus failure modes
- Voice of the customer versus failure modes
- Quality assurance planning by part
- Supplier versus manufacturing parts/materials
- Process failure analysis

The entire system encompasses several phases of product development, with a strong continuous improvement emphasis. The phases have been labeled as follows:

Matrix	"WHAT"	"HOW"	Activity
A1	Voice of Customer	SQCs	Construct Matrix
A2	Functions	SQCs	Construct Matrix
A3	SQCs	SQCs	Construct Matrix
A4	Second level of design	SQCs	Construct Matrix
B1	Voice of Customer	Functions	Construct Matrix
B2	Competitive analysis	Cost	Construct Matrix
B3	Detailed SQCs	Breakthrough targets	Construct Matrix
B4	Critical parts	SQCs	Construct Matrix
C1	New technology	First level of design	Construct Matrix
C2	Functions	First level of design	Construct Matrix
C3	SQCs	First level of design	Construct Matrix
C4	Second level of design	First level of design	Construct Matrix
D1	Voice of Customer	Product failure modes	Construct Matrix
D2	Functions	Product failure modes	Construct Matrix
D3	SQCs	Product failure modes	Construct Matrix
D4	Second level of design	Product failure modes	Construct Matrix
E1	Customer needs	New concepts	Construct Matrix
E2	Functions	New concepts	Construct Matrix
E3	SQCs	New concepts	Construct Matrix
E4	Criteria	New concepts	Construct Matrix
F1			Value engineering
F2			Reliability analysis
F3			Breakthrough planning
F4			Design improvement planning
G1			Quality assurance planning
G2			Equipment deployment
G3			Process planning
G4			Process FTA
G5			Process FMEA
G6			Process QC

Diagram 18-5. Matrix of Matrices Summary

- Voice of the Customer phase
- Define product and identify bottlenecks and breakthrough opportunities
- Develop design breakthroughs
- Develop process breakthroughs

The Akao QFD model is not intended to be used verbatim. It is intended to open up possibilities to a development team. The team is expected to create its own QFD model, because no two organizations and no two development projects have the same needs.

As indicated in Section 1.2, the Akao model was introduced in the U.S. by Bob King around 1984. In the early nineties, King and others proposed a subset of the thirty matrices for those developers who were overwhelmed by the matrix of matrices—so overwhelmed that they were unable to realize any benefit from QFD at all.

The subset, called "Comprehensive QFD," included seventeen matrices generally based on the original thirty, and it added the Voice of the Customer Table and a concept selection activity. The matrices were presented within a grid that had product development phases along the left (moving from VOC acquisition and proceeding to detailed design, as in the Four-Phase Model), and the elements of continuous improvement ("Plan–Do–Check–Act") along the top. "Plan–Do–Check–Act" refers to the Deming/Shewhart cycle of continuous improvement,[6] which is a fundamental TQM concept. Thus, this formulation of QFD creates a continuous improvement context for product development. The elements of "Comprehensive QFD" are listed in Diagram 18-6.

The use of "Plan–Do–Check–Act" for structuring the quality tables is consistent with the notion that strategic corporate quality planning, the PDCA con-

Activity	Matrix
Marketing/Planning	Voice of Customer Tables
System-Level Design/Planning	VOC/SQCs VOC/Functions SQCs/Functions
System-Level Design/Checking	Key SQCs/Methodology Key Functions/Methodology Key VOC/Failure Modes Key Functions/Failure Modes
Concept-Level Design/Doing	Key SQCs/Concepts Key SQCs/Top Concepts
Concept-Level Design/Checking	Key Concepts/Methodology Key Concepts/Failure Modes Key Concepts/Cost Key Concepts/Safety
Concept-Level Design/Acting	Select Best Concepts
Element-Level Design/Doing	Key SQCs/Elements of Best Concept
Element-Level Design/Checking	Key Elements/Methodology Key Elements/Element Failure Modes Key Elements/Cost

Diagram 18-6. Comprehensive QFD Matrices (Summary)

tinuous improvement model, and QFD can be combined to provide an overall TQM model.

Summary

In this chapter we have looked at the Four-Phase Model and the Matrix of Matrices Model of QFD. We've compared them and concluded that the more elaborate model, the Matrix of Matrices, makes explicit activities that are implicit or optional in the Four-Phase Model. Otherwise, the two models are not in conflict.

The Four-Phase Model identifies a series of translations from the Voice of the Customer through to production process steps. The Matrix of Matrices does the same, but also makes explicit various forms of reliability engineering, cost analysis, value engineering, and quality control.

QFD is a system of matrices that provides a backbone for the entire development process. Various models of QFD as a development backbone exist. The best known of these are the ones described in this chapter (the Four-Phase Model and the Matrix of Matrices.) Neither model is intended to be used blindly as presented. Instead they are intended to provide templates or examples of the possibilities for systematic planning throughout the development cycle. Both models suggest that the Voice of the Customer can be carried through the development cycle to the very details of implementation. The more elaborate models suggest that notions of continuous improvement can be integrated with QFD.

We mentioned in the introduction to this book, and in many other places, that QFD helps any team relate its "Whats" to its "Hows." A great deal of creativity has already been applied to extending QFD's initial purpose of product development. In the next chapter, we'll look at some ingenious ways that QFD has been used. Perhaps readers of this book will be inspired to apply QFD in yet other creative ways.

Discussion Questions

What does "Plan–Do–Check–Act" mean in your organization? Is it integrated with product or service development? Can you conceive of a multilevel QFD structure that would support and improve your development process? What would it look like? What do your colleagues think of it?

---

[1] Andreas Krinninger and Don P. Clausing, "Quality Function Deployment for Production," Working Paper published by the Laboratory for Manufacturing and Productivity of the Massachusetts Institute of Technology, Cambridge, Massachusetts, August 1991.

[2] Dr. Don Clausing, *Total Quality Development*, ASME Press, 1994.

[3] Ibid.

[4] Andreas Krinninger and Don P. Clausing, "Quality Function Deployment for Production," Working Paper published by the Laboratory for Manufacturing and Productivity of the Massachusetts Institute of Technology, Cambridge, Massachusetts, August 1991.

[5] Bob King, *Better Designs in Half the Time*, GOAL/QPC, Methuen, Massachusetts, 1987.

[6] W. Edwards Deming, "Quality, Productivity, and Competitive Position," MIT Center for Advanced Engineering Study, Cambridge, Massachusetts 02139, 1982.

CHAPTER 19

Special Applications of QFD

Introduction

In this chapter we'll look at some of the ways QFD has been extended beyond its initial conception. The specific areas are QFD for Total Quality Management, Strategic Product Planning, Organizational Planning, Cost Deployment, Software, and Service.

QFD's primary application has been for planning and managing product development. The dominant QFD models assume hardware product development activities. As we have seen, however, the QFD model can easily be applied to development of almost any type of product or service. My own initial experiences with QFD were in the area of software development. Very quickly afterward, I used QFD to help an internal service group develop its strategy for customer satisfaction of its internal customers.

Moreover, the general idea of matching "Whats" against "Hows" to understand their linkage and importance has even wider application than product development. Let's begin our survey of these newer applications by considering QFD and Total Quality Management.

320

19.1 Total Quality Management

There are many different definitions of TQM. All approaches to TQM contain a strong element of aiming for customer satisfaction. Putting this into QFD terminology, a common pursuit in TQM is finding a cost-effective way to link or align the organization's activities to best meet the needs of the customer.

One very powerful and widely understood formulation of this is the application criteria for the Malcolm Baldrige National Quality Award. The Baldrige criteria define seven major aspects of TQM. The exact detailed formulation of these seven aspects varies slightly from year to year, but the general areas haven't changed since the award's inception.

The seven areas are these:
1. Leadership
2. Information and Analysis
3. Strategic Quality Planning
4. Human Resource Development and Management
5. Management of Process Quality
6. Quality and Operational Results
7. Customer Focus and Satisfaction

In any TQM model, these general areas must be linked to customer needs and to each other to form a coherent program. Some of the detailed questions Baldrige applicants must address in their written applications are these (paraphrased):
- How are the company's quality objectives translated into work units' operations plans?
- How are quality values integrated into the company's management system?
- What is the approach for each type of competitive comparison and benchmarking?
- How are performance-related results and quality feedback analyzed and translated into actionable information for developing priorities in operations?
- How does the planning process create a framework for customer satisfaction and leadership?
- How does the planning process drive improvement in operations and processes?
- How are plans communicated to work units and suppliers?

Other approaches to TQM, such as Hoshin Kanri, Policy Deployment, and Reengineering, all call for linkages between customer needs and the organization's functioning.

While the Baldrige criteria prescribe no answers for such questions, and while there may be other ways to establish the linkages, more and more U.S. companies are now using QFD for this purpose. Customer needs are formulated at a very high level and deployed to corporate objectives. These objectives are deployed through the levels of the corporation (and the number of levels is being reduced in many of these companies). QFD is becoming the vehicle of choice for these activities.

19.2 Strategic Product Planning

19.2.1 Planning for Families of Products

Many business groups sell a family of products into a market. Examples are

- Systems of computers and software
- Families of chemicals or drugs
- Telephones and accessories
- Financial services

In order to obtain a better understanding of how the family of products work together to meet the market's needs, some management teams are using QFD. A typical approach is shown in Diagram 19-1.

The planning matrix is used for strategic goal setting. The impacts of the product family members on meeting needs determines the priorities of the product family members. The impacts also indicate whether certain products are redundant, and whether certain needs are not being properly met.

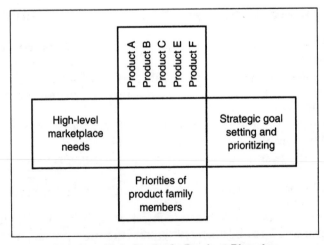

Diagram 19-1. Strategic Product Planning

19.2.2 Planning Multiple Products for Multiple Market Segments

Background

A large company recently embarked on a new product strategy. In the past, they had developed a limited number of product lines. They had been using a flexible pricing strategy to sell the same products into various markets. They recognized, however, that by lowering their prices to cost-conscious customers, they were lowering margins substantially, and could not stay competitive indefinitely. In effect, most of their products were too rugged and durable to allow them to compete in cost-conscious markets.

The company employed VOCALYST (Section 16.5) to acquire the VOC for four different *types* of customers or users, all of whom were involved in purchasing, installing, or using a product. The types were

- **Recommenders:** People who recommended products for purchase, such as consultants
- **Managers:** People who made purchase decisions
- **Maintainers:** People who installed the products and provided ongoing maintenance of them
- **End Users:** People who actually used the products on a day-to-day basis

Note that these types do not represent alternative market segments into which the same product would be sold. Rather, they are all in the same market, and even represent the same customer. They simply have different perspectives and needs for the product, at different points in the product's life. The partial list of secondary and tertiary customer needs (attributes) looked similar to those in Diagram 19-2.

Using the Voice of the Customer to Develop a "Master House of Quality"

The development team that used the VOC data was a ten-person, high-level multifunctional team, with members representing marketing, product design, sales, and manufacturing functions. The development team recognized that the strategic direction of their company lay in positioning their products at much smaller and more uniquely defined market segments than they had hitherto identified. In this way the product capabilities and pricing could be better targeted, more competitive, and therefore more profitable for each market segment. Even at the time of the market study, however, these market segments had not been identified. Furthermore, it was expected that new market segments would continue to be identified and targeted in the foreseeable future.

Because the strategic market segments were to be defined on an ongoing basis, the team could not predict the relative importance of the customer needs for

Secondary Attribute	Tertiary Attribute
Easy to Order	• Can get options I want quickly • Can understand prices easily • Equipment arrives in good condition
Easy to Explain	• Simple, intuitive controls • Controls within easy reach • Can learn adjustments without a manual
No Data Loss	• Data is safe even when there is a power failure • System continues to run even if a component fails • Can always retrace my processing steps

Diagram 19-2. Examples of Affinity Groupings

any particular market segment, although they were confident that the needs themselves would not change much over time. Therefore, although the VOC-ALYST process delivered both qualitative and quantitative data about the four customer types, the team felt it could only use the qualitative data, i.e., the tertiary and secondary customer attributes. They felt comfortable that the Voice of the Customer process had uncovered "universal" secondaries that constituted a superset of the secondaries needed to describe each as-yet-undefined market segment.

The team's strategy for using QFD was then to create a "master" HOQ that listed all secondaries for all four customer types on the left. A successful product would have to satisfy all four customer types, regardless of the market segment. However, the relative importance of the four customer types, and the relative importance of each secondary attribute within the customer type, was expected to vary with the market segment. The team decided that the company would have to determine the relative importances of the needs in the future for each market segment.

Accounting for some duplicate secondaries shared among different customer types, the team ended up with a list of approximately fifty secondary attributes. They then generated performance measures (Substitute Quality Characteristics) as the column headings in the House of Quality for each secondary customer need. As is common in generating SQCs, two or three performance measures were required for each secondary customer need. The result was approximately 150 performance measures. This resulted in a Relationship section of about 7500 cells—a formidable HOQ by any standards.

Nevertheless, the team committed itself to filling in the entire matrix. The result was a master HOQ that defined customer needs, performance measures, and their linkages applicable to many market segments, far into the future. Given that the secondary customer attributes, the performance measures, and their linkage to customer attributes were universal, the only missing data, to be supplied by individual development teams for their market segment in the future, were the relative importance values of the four customer types, and the relative importance of the secondary customer needs.

Next, the team created a handbook that explained to future product developers in the company how to customize the master HOQ for their specific market segments. Each future product team will determine the relative importance of the four customer types and the secondary customer attributes for *their* market (an example is shown in Diagram 19-3). The team will then plug those numbers into the master HOQ, and thereby prioritize the performance measures for their market segment.

By experimenting with some hypothetical importance values, the team determined that it would be possible for most future development teams to identify a small handful of performance measures out of the master list of 150 that would be critical for success in a particular market segment. Thus the master House of Quality will be the source for many market-specific Houses of Quality that will be created by future market segment teams. Schematics of such market-specific Houses of Quality are shown in Diagram 19-4.

The process was highly publicized throughout the company, and it generated a great deal of interest. Even before the master HOQ was completed, a few development teams had begun using the preliminary results.

This case study demonstrates the use of an excellent VOC technique, combined with QFD, to create a strategic product development methodology that will influence product development for years. The resulting master HOQ provides a company-wide statement of both customer needs and top-level corporate response to meeting those needs that can be customized by product development teams throughout the organization.

19.3 Organizational Planning

19.3.1 Selecting an Organizational Schema

Often (perhaps too often!) it becomes necessary to change the structure of all or part of a company's organization. Many mutually exclusive choices must be made, such as:

> How many new organizations will replace the old?
> Who should manage each new organization?

Customer Segment & Segment Importance	Secondary Attribute	Importance Within Segment	Overall Importance
Recommenders .40	Easy to order	100	40
	Easy to explain	70	28
	Parts available	60	24
	On-time delivery	60	24
Managers .30	Durable products	100	30
	Good looking	50	15
	No safety hazards	90	27
Maintainers .15	No data loss	100	15
	Easy cabling	70	11
	Protected from damage	60	9
End Users .15	Equip. at right height	80	12
	Reliable power source	60	9
	Organize work easily	100	15

Diagram 19-3. Customer Types, Attributes, and Importance Values for a Market Segment

How should organizational objectives be assigned?
How will people be reassigned from the old organization to the new organization?

There may be several plausible options for each of these choices, making the combinations of plausible sets of options difficult to choose from.

A combinatorial approach combined with the Pugh concept selection process can help.

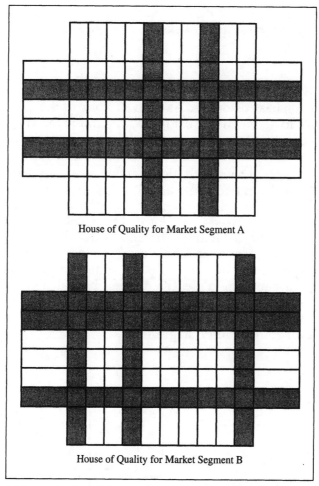

House of Quality for Market Segment A

House of Quality for Market Segment B

Diagram 19-4. Examples of Market-Specific Houses of Quality

The most promising combinations of answers, plus today's organization, are selected as alternative concepts and labeled: "Today's org." and Proposals A, B, and C. "Today's org." is selected as the datum. The datum and the proposals are thus the initial concepts and are shown above the double horizontal line in Diagram 19-5.

The objectives it is hoped that the reorganization will achieve are placed on the left of matrix. They are the criteria that management hopes to optimize. The proposals are compared to the datum. As the management team evaluates the proposals against the datum, they will discover advantages and disadvantages, just as any in any other well-run concept selection exercise.

The possibility of hybrids and other variations of the initial concepts exists, and the opportunity for finding the best plan is enhanced.

19.3.2   Matching the Organization's Work to Its Objectives

Every organization has customers. As with all Voice of the Customer activities, it is possible to define these customers, determine the attributes of customer satisfaction, and measure current satisfaction performance levels.

These organizations also have activities that they normally perform. QFD can be used to evaluate the impact these activities have on customer satisfaction, as shown in Diagram 19-6. In this diagram the entries on the left of the matrix are examples of customer needs (the Voice of the Customer). The processes along the top are Substitute Quality Characteristics. They are a list of the organization's primary functions. These can be determined either by the organization's strategic plan, or by a team brainstorming process. In the case of brainstorming, the team would identify all the jobs that anyone in the organization does, and then develop an affinity diagram of all jobs brainstormed. The processes at the top of the matrix in Diagram 19-6 would be the primary or secondary level of the affinity diagram.

	Today's organization	Proposal A	Proposal B	Proposal C
Number of organizations	3	4	2	1
Who should manage?	L.C.	R.L.	R.L.	J.O.
Organizational objectives	Plan A	Plan B	Plan C	Plan C
Reassignment plan	Plan x	Plan y	Plan x	Plan y
Better customer service	D	+	-	-
Reduce organizational levels	A T U M	-	-	+
Reduce head count		S	+	S
Higher morale		S	-	+

Diagram 19-5. Organizational Concept Selection

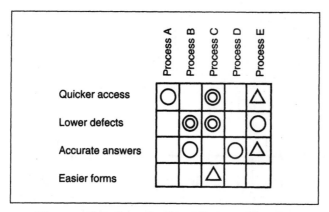

Diagram 19-6. Organizational Process Planning

Once the Relationships have been evaluated, the team will understand how each of their processes affect their customer's needs. They will then see which processes have the greatest impact on meeting the needs. Thus, they can decide which processes must be delivered most efficiently.

19.4   Cost Deployment

Cost deployment is used to allocate known development costs of any type to the customer needs or "Whats" they support. Cost deployment can show a team what they are paying to support each function or customer need. The team can determine whether they are paying too much to support unimportant needs, and therefore whether new concepts are needed.

The costs of the "Hows" are arrayed along the top of the Cost Deployment matrix. In Diagram 19-7, a tree diagram representing a product design is shown at the top of the matrix. The product parts are at the lowest level of the tree, and below each part is the cost of that part.

The team places the "Whats" along the left side. In Diagram 19-7, the "Whats" are customer needs. The team then estimates the contribution of each "How" to each "What" as a fraction. In the diagram these are the numbers below the diagonal line in each cell. The fractions in each column should add to one. The team multiplies the cost of each "How" by the fractions and puts the resulting contribution in the same cell (above the diagonal line). The contribution represents the estimated cost that the "How" contributes to meeting the customer need.

The sum of the contributions in a row is the total cost to meet the "What" represented by that row. The total costs can be compared to the raw weights from

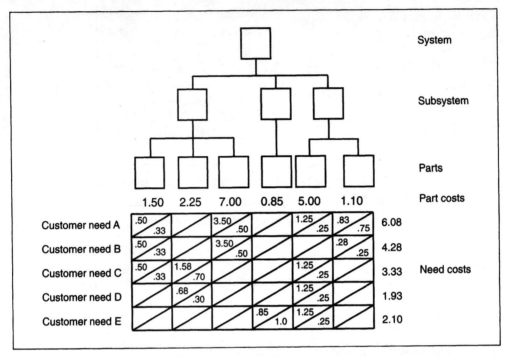

Diagram 19-7. Cost Deployment

the planning matrix, or to the priorities of these "Whats" as the result of any other matrix analysis. A serious misalignment of costs to priorities is a good indication of the need to redesign.

In the development of complex products such as computer systems, the costs of final testing are usually very high. Cost deployment is an excellent way of determining whether the cost of each test is appropriate to its purpose. The Voice of the Customer is placed on the left of the matrix, and the final tests are placed at the top. The fractional values below the diagonals are the extent that the test contributes to verifying that the product meets the customer satisfaction performance targets. The cost of each test is at the top of each column. The completed matrix can then indicate the total cost of verifying success in meeting each need.

19.5   Software Development

The Four-Phase QFD model is based on the paradigms of designing and manufacturing physical objects—hardware. Software paradigms are different enough that software engineers cannot use this model beyond the House of

Quality. In fact, most QFD applications for hardware, software, and services use only the House of Quality.

The author[1-3] has applied QFD to software development by making extensive use of the House of Quality, using software functions as the Substitute Quality Characteristics, and deploying these functions down more than one level.

Two important software analysis and design paradigms have been reconciled with QFD. Richard Zultner[4] has developed a QFD model that incorporates elements of the "structured analysis" paradigm. Professor Hisakazu Shindo has developed a QFD model that incorporates elements of object-oriented analysis and design.[5]

19.5.1   Structured Analysis and QFD

Structured analysis is a general term referring to a number of graphical techniques aimed at helping software analysts translate customer requirements into technical requirements. Several structured analysis systems have been developed since the early seventies.[6,7] In general, these methods transform the technical requirements into data elements and procedural elements.

In Zultner's QFD model, the Voice of the Customer is deployed into technical requirements, the highest level of Substitute Quality Characteristics. The language of technical requirements refers to high-level functions of the software product, such as "provide graphic editing capability."

The technical requirements are then deployed into two separate sets of lower-level SQCs: data elements and procedural elements, as in Diagram 19-8. Procedural elements might include "provide editing of circles" and "provide editing of rectangles." Data elements might include "circle" and "rectangle."

The procedural elements or processes and the data elements are prioritized independently against the technical requirements. These prioritizations can then be used to guide in resource allocation, technical breakthrough focusing, and testing resources, for example.

19.5.2   Object-Oriented Analysis and QFD

Object-oriented analysis and design brings procedures and data together into objects. A software object is a description of some physical or conceptual object, together with the procedures that apply to that object. For example, the object "circle" might include the data that represents the circle, along with the operations that the software can perform on a circle, such as CREATE, MOVE, CHANGE SIZE, and DELETE.

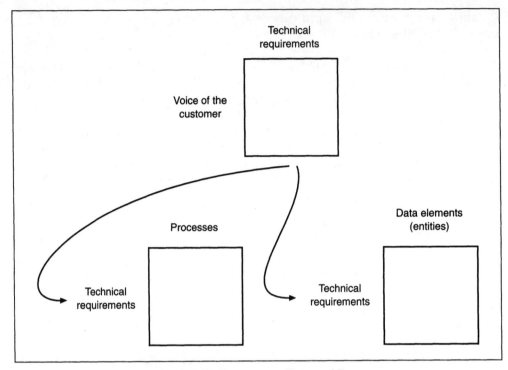

Diagram 19-8. Deployment to Data and Processes

A primary advantage of objects is that they can be conveniently reused and modified. They are normally placed in large "class libraries" that software engineers can select from. However, defining the objects incorrectly renders them difficult to use and reuse. One of the challenges of object-oriented analysis and design is to "discover the objects." Defining just what data and procedures should be encapsulated into a single object has proven to be difficult.

In Shindo's approach, the Voice of the Customer is deployed into Technical Requirements. The technical requirements are deployed into processes and data elements. Then the processes and data elements are placed on the left and top of a new matrix (Diagram 19-9). The matrix is input to a matrix reduction and manipulation process known as Quantitative Method Type III, which interchanges rows and columns to best arrange the nonzero matrix elements along the diagonal. This gathers those processes and data together that would appear to be best encapsulated into objects.

While Quantitative Method Type III provides an automatic method or clustering processes and data, it is possible to use judgment and observation to perform the process without automation.

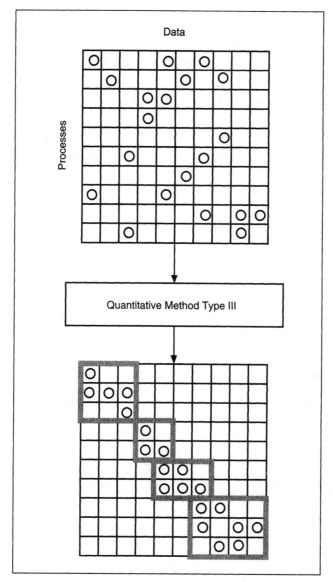

Diagram 19-9. Quantitative Method Type III

19.6 QFD for the Service Industry

QFD has been used in the service industry for quite a while. References to service QFD date from as early as 1988 in Japan, in Yoji Akao's collection of Japanese papers on QFD.[8] This author used QFD for service applications in 1987, without having seen any published case studies for service QFD.

Step ID	What	Where	Who	When	Why
1	Ask customer how you can be of service	Customer counter	Customer Service Rep.	Business hours, 8:30 A.M. to 5:00 P.M.	
2	If customer has existing account, get account number	Customer counter	Customer Service Rep.	Business hours, 8:30 A.M. to 5:00 P.M.	Verifies customer really has account, brings up account information on terminal
3	If customer wants account balance, get balance from terminal, write it on "Acct. Balance" form and hand it to customer	Customer counter	Customer Service Rep.	Business hours, 8:30 A.M. to 5:00 P.M.	By writing balance instead of saying it, customer is given privacy

Diagram 19-10. Detailed Process Steps

A good formulation for service QFD helps the development team identify the detailed process steps required to perform a service. As described in Section 7.3, performance measures or process functions can be used as SQCs. If performance measures are used, then deployment of those measures to processes or subprocesses is a logical next step. Once deployed to the process or subprocess step, some teams are now developing detailed process steps, using a "What/Where/Who/When/Why" format that resembles Diagram 19-10.

Once the information is in this format, and especially if the format is input to a computer database, the steps can be sorted by the person responsible, by time, or by any of the columns, to manage responsibilities, timing, and even space.

Summary

In this chapter we have seen how QFD can be used for many applications far beyond the initial concepts for QFD. The application areas ranged from Total Quality Management to object-oriented design of software.

The better we understand QFD, the easier it is to see how to use it in new situations. The inescapable conclusion is that QFD is a very robust tool with as yet unbounded application. Suppose we restrict ourselves to the planning and development of products and services. We need only look around us at products that don't work well or services that aren't worth the money we paid for them, to realize how many more traditional applications of QFD are needed.

I encourage you to think of products and services you use that might benefit from QFD. Then consider products or services that you are responsible for delivering. Could QFD help to make them better?

Beyond that, we can hope to see systematic analysis of "Whats" and "Hows" applied in many untapped areas.

Discussion Questions

How could or should QFD be applied to the following: Personal investment strategy? Selection of a new car? Deciding who to vote for in an election? The process of legislation? Formulation of public policy? International diplomacy?

[1] Lou Cohen, "QFD: The House of Quality," *American Programmer,* June 1993.

[2] Lou Cohen, "Quality Function Deployment: An Application Perspective from Digital Equipment Corporation," *National Productivity Review,* Summer 1988.

[3] Lou Cohen, "Insights into QFD at Digital Equipment Corporation," Proceedings of National Electronic Packaging and Production Conference, June 1992.

[4] Richard Zultner, "Quality Function Deployment for Software," *American Programmer Magazine,* February 1992.

[5] An English-language description of this approach can be found in Takami Kihara and Malik Mamdani, "Quality Function Deployment Applied to Software at Digital Equipment Corporation," working paper at the Thayer School of Engineering, Dartmouth College, May 1991.

[6] Edward Yourdon and Larry L. Constantine, *Structured Design: Fundamentals of a Discipline of Computer Program and System Design,* Yourdon Press, A Prentice-Hall Company, 1979.

[7] Chris Gane and Trish Sarson, "Structured Systems Analysis: Tools and Techniques," MCAUTO McDonnell Douglas, 1982.

[8] Teiichiro Noda and Junji Ogino, "Quality Function Deployment for the Service Industry," translated into English in *Quality Function Deployment: Integrating Customer Requirements into Product Design,* Yoji Akao, editor, Productivity Press, 1990. This collection of papers is a translation of "Hinshitsu tenkai katsuyo no jisai," published by Japan Standards Association, 1988.

Afterword

We have come a long way in this book. The steps we've described take us from customer's needs to delivered product or service. One might say that they all make sense, and many development groups perform some of them already. What makes the QFD approach different from previous development methods is the systematic analysis of the relationships between the "Whats" and "Hows," and the invitation to envoke this systematic analysis at many different planning points in the development process.

I have had the privilege of working with many development groups, for twenty years before I ever heard of QFD, and for close to ten years since. There is no question in my mind that QFD is a robust structure that provides purpose and logic to the product-planning process, an area that had previously stubbornly remained an unstructured "gray area" to me and my colleagues.

Before QFD, the development of product requirements was an arcane activity entrusted to a few gifted technical leaders. The proof of their ability to meet customer needs was something that only time would tell, after the product had been built and sold.

Now that I have learned about QFD, I see the development of product requirements as a clear series of steps that teams can follow. Decisions can be backed up with documented judgments based on a clear understanding of the customer.

It's true that QFD, properly done, takes time. What is often overlooked is that product planning has always taken time. The debates and controversies that raged in reaction to someone's product proposal were highly emotional. Arguments were posed without substantive basis, and nothing got decided. Organizations were paralyzed, and projected delivery dates slipped. All of that took excessive amounts of time. We couldn't afford it in the past, and it is even more intolerable today.

337

Nowadays, many global companies have learned to match customer needs to development details in time periods short enough to have taken my breath away fifteen years ago. Any organization that wants to compete today must compete with these high-performing global companies. No one can afford to develop products or services that do not meet customer needs the moment they are delivered.

I have tried to present QFD as a flexible tool that can be fashioned to be effective in a wide range of applications and for many types of organizations. I hope the reader will have grasped the essentials of using QFD so that it will be easy to modify the process without sacrificing those advantages that will be important for a given project.

Up to now, if you have been reading this book, you will have been learning about QFD and how it supports development. I've given you what I know of the topic. Now it's time for you to use QFD in your pursuit of producing great products and services. My best wishes to you.

Bibliography

This bibliography lists those books that have had an important influence on my knowledge of quality management and QFD, as well as all English-language books on QFD that I am aware of. My apologies to any author whose book is not mentioned here; I have done my best to be all-inclusive.

Yoji Akao, Editor-in-Chief, *Quality Function Deployment: Integrating Customer Requirements into Product Design,* translated by Glenn H. Mazur and Japan Business Consultants, Ltd., Productivity Press, Cambridge, Massachusetts, 1990.

James L. Bossert, *Quality Function Deployment: A Practitioner's Approach,* ASQC Press, 1991.

Michael Brassard, "Memory Jogger Plus+II," GOAL/QPC, 1994.

Don Clausing, *Total Quality Development,* ASME Press, 1994.

Ronald Day, *Quality Function Deployment: Linking a Company with Its Customers,* ASQC Quality Press, 1993.

W. Edwards Deming, *Out of the Crisis,* Massachusetts Institute of Technology, Center for Advanced Engineering Study, Cambridge, Massachusetts, 1992.

W. Edwards Deming, *Quality, Productivity, and Competitive Position,* Massachusetts Institute of Technology, Center for Advanced Engineering Study, Cambridge, Massachusetts, 1982.

Abbie Griffin, "Functionally Integrating New Product Development," Ph.D. Thesis, MIT, 1989.

Abbie Griffin and John Hauser, "The Voice of the Customer," Report Number 92-106, Marketing Science Institute, Cambridge, Massachusetts, March 1992.

J. M. Juran and Frank M. Gryna, Jr., *Quality Planning and Analysis,* McGraw-Hill Book Company, 1980.

Bob King, *Better Designs in Half the Time,* GOAL/QPC, Methuen, Massachusetts, 1987.

S. Marsh, J. W. Moran, S. Nakui, and G. Hoffherr, *Facilitating and Training in Quality Function Deployment*, GOAL/QPC, 1991.

Shigeru Mizuno, Editor: *Management for Quality Improvement: The Seven New QC Tools*, Productivity Press, 1988.

Novi Annual QFD Symposium Transactions, *Transactions from the Symposium on Quality Function Deployment*, 1989 to the present. The symposium has been jointly sponsored by the American Supplier Institute, Inc., and GOAL/QPC.

Amitabh Pandey, "Quality Function Deployment: A Study of Implementation and Enhancements," MIT Master's Thesis, 1992.

Madhav S. Phadke, *Quality Engineering Using Robust Design*, Prentice-Hall, Englewood Cliffs, New Jersey, 1989.

Stuart Pugh, *Total Design*, Addison-Wesley, Reading, Massachusetts, 1991.

Thomas L. Saaty, "Decision Making for Leaders" and "The Analytical Hierarchy Process," both published by RWS Publications, 4922 Ellsworth Avenue, Pittsburgh, Pennsylvania 15213, 1990.

Shoji Shiba, Alan Graham, and David Walden, "A New American TQM: Four Practical Revolutions in Management," Productivity Press/Center for Quality Management, 1993.

Genichi Taguchi (Don Clausing, technical editor for the English edition), *System of Experimental Design*, UNIPUB/Kraus International Publications, 1987 (2 vols.).

Glen Urban and John Hauser, *Design and Marketing of New Products*, 2nd edition, Prentice-Hall, Englewood Cliffs, New Jersey, 1993.

John N. Warfield, *A Science of Generic Design: Managing Complexity through System Design*, second edition, Iowa State University Press, 1994.

Index